Real-Time Heterogeneous Video Transcoding for Low-Power Applications

Tarek Elarabi · Ahmed Abdelgawad
Magdy Bayoumi

Real-Time Heterogeneous Video Transcoding for Low-Power Applications

Tarek Elarabi
Intel
Santa Clara, CA
USA

Magdy Bayoumi
University of Louisiana at Lafayette
Lafayette, LA
USA

Ahmed Abdelgawad
Central Michigan University
Mount Pleasant, MI
USA

ISBN 978-3-319-38131-2 ISBN 978-3-319-06071-2 (eBook)
DOI 10.1007/978-3-319-06071-2
Springer Cham Heidelberg New York Dordrecht London

Printed on acid-free paper

Springer is part of Springer Science+Business Media (www.springer.com)

I lovingly dedicate this book to my wife,
Dr. Aumnia Farag
who supported me each step of the way.
There is no doubt in my mind that without her continued support and counsel, I could not have completed this work

Tarek Elarabi

Acknowledgments

My deepest gratitude is to Professor Madgy Bayoumi. I have been amazingly fortunate to have an advisor who gave me the freedom to explore on my own, and at the same time the guidance to recover, when my steps faltered. Dr. Bayoumi taught me how to question thoughts and express ideas. His patience and support helped me overcome many crisis situations and finish this book. I hope that one day, I will become as good an advisor to my students, as Dr. Bayoumi has been to me.

I would like to show my greatest appreciation to Ryan Ahrabi. I cannot say thank you enough for his tremendous support and help. I feel motivated and encouraged every time I attend his meeting. I am grateful for his constant support and help.

I thank Dr. Hongyi Wu and Dr. Nian-Feng Tzeng for their support, guidance, and suggestions as I moved from an idea to a completed study. I am also indebted to the faculty and staff of the Center for Advanced Computer Studies, with whom I have interacted during the course of my graduate studies. I am heartily thankful to Dr. Chu Chee-Hung Henry, whose encouragement, advice, guidance, and support from the initial to the final level, enabled me to develop an understanding of the subject. I owe my cordial thanks to Dr. Jason McNeely for numerous discussions we have shared and for providing support and honest feedback on my work on the video technology.

I also would like to acknowledge my wife, Aumnia and my son, Yousef for supporting and encouraging me to pursue my doctorate. Without my wife's encouragement and help, I would not have finished this book. I am also thankful to her for encouraging the use of correct grammar and consistent notation in my writings and for carefully reading and commenting on countless revisions of this manuscript.

Finally, I offer my regards and blessings to all those who supported me, in any respect, during the journey of my engineering education.

Tarek Elarabi

Contents

Acronyms

AC	The Arithmetic Coding
ASIC	Application Specific Integration Circuit
ASP	Advanced Simple Profile
AVP	Advanced Visual Profile
BP	Baseline Profile
CABAC	Context Adaptive Binary Arithmetic Coding
CCT	Conventional Cascaded Transcoder
CDP	Chroma Direction Prediction
Chroma	Chrominance
CIF	Common Intermediate Format
CPU	Central Processing Unit
DCT^{-1}	Inverse Discrete Cosine Transform
DTV	Digital TV
DVB	Digital Video Broadcasting
EP	Extended Profile
FPGA	Field-Programmable Gate Array
FSF	Full-search Free
HD	High-Definition
HDL	Hardware Description Language
iDCT	Discrete Cosine Transform
ISO/IEC	International Organization for Standardisation/International Electro-technical Commission
ITU-T	International Telecommunication Union—Telecommunication Standardization Sector
JM 18.2	H.264/AVC reference software
LCF	Lagrangian Cost Function
Luma	Luminance
MB	Macro-block
MD	Mode Decision
MP	Main Profile
MPEG	Moving Picture Experts Group
MVC	Multi-view Video Coding
NH	Non-Homogeneity
PSNR	Peak Signal-to-Noise Ratio

Q^{-1}	Inverse quantization process
QP	Quantization Parameter
RD	Rate Distortion
RDO	Rate Distortion Operations
S	Smoothness factor
SAD	Sum of Absolute Differences
SATD	Sum of Absolute Transformed Differences
SD	Standard-Definition
SHP	Stereo High Profile
SSD	Sum of Square Differences
T^{-1}	Inverse transformation process
Th	Threshold
VCEG	Video Coding Experts Group
VHDL	Very high speed integrated circuits Hardware Description Language
VLC	Variable Length Coding

Chapter 1
Introduction

Abstract In this book, we have presented a high throughput, yet precise, MPEG-2 to H.264/AVC transcoding algorithm that achieves an average of 92.8 % reduction in the required transcoding run-time at a price of an acceptable PSNR degradation, which nominates our proposed transcoding solution for real-time video applications. In particular, the proposed accelerating MPEG-2 to H.264/AVC transcoding algorithm employs a novel Intra mode selection and direction prediction technique that utilizes the DCT coefficients directly from the MPEG-2 side of the transcoder for such purpose. The proposed Intra prediction methodology successfully decides the Intra macro-block mode, then predicts the corresponding reconstruction direction for the H.264/AVC encoder side of the transcoder in a completely Full-Search Free technique. In fact, the H.264/AVC standard only documents the Intra mode decision and direction prediction process for the decoder version on the H.264/AVC codec. For such reason, we modified our proposed transcoding Intra prediction algorithm to fit as an enhancement to the H.264/AVC standard decoder. For this purpose, also the properties of the DCT coefficients are employed to significantly reduce the overall run-time of the H.264/AVC standard decoder. Specifically, a high throughput early Intra mode decision and direction prediction algorithm is presented in the scope of this book, which has been entitled "FSF Intra Prediction algorithm". The algorithm also utilizes the macro-block DCT coefficients directly from the output of the inverse quantization module for such purpose and before macro-blocks' samples get transformed back to the pixel domain. The novelty of the proposed FSF algorithm is that it selects the macro-block Intra mode, then predicts the macro-block reconstruction direction not only fast, but also in a completely Full-Search Free environment. The experimental results show that the FSF Intra prediction software implementation achieves over 56 % reduction in the Intra prediction run-time while preserving the same bit-rate and subjective quality as the standard H.264/AVC. However, the cost of the obtained run-time reduction is a negligible PSNR degradation of less than 0.72 dB when it is compared to the PSNR achieved by Intra prediction algorithm that currently employed by the H.264/AVC JM 18.2 reference software. Moreover, we extended our work by introducing the low-power hardware architecture design for the

T. Elarabi et al., *Real-Time Heterogeneous Video Transcoding for Low-Power Applications*, DOI: 10.1007/978-3-319-06071-2_1,

proposed high-throughput FSF Intra prediction algorithm. Furthermore, both FPGA prototype and ASIC chip design and implementation are developed for the proposed FSF Intra prediction hardware architecture. In closing, the proposed 45 nm ASIC design achieves a stable operating frequency of 140 MHz, which strongly nominates the FSF Intra prediction hardware design for real-time H.264/AVC video devices. Even more, the total power consumption of the ASCI design was only 9.1 mW, which qualifies the proposed chip design for low-power mobile video applications.

1.1 The H.264/AVC Standard in Glance

H.264/AVC, the video coding standard recommendation of ITU-T (International Telecommunication Union—Telecommunication Standardization Sector), is jointly developed and co-published with ISO/IEC (International Organization for Standardisation/International Electro-technical Commission). The joint efforts of both video groups have paid off the H.264/AVC, which is a breakthrough in the area of video coding.

By finalizing the H.264/AVC video standard, they have succeeded to achieve more than double the compression rate while preserving the visual video quality. Moreover, the H.264/AVC, which is also known as MPEG-4 part 10, has been widely implemented in most mobile multimedia terminals because of its high encoding efficiency. It guaranteed over 50 % bit rate reduction compared to former video standards while preserving minimum quality degradation.

As illustrated by Fig. 1.1 [1], a comparison between the coding efficiency of the H.264/AVC and the MPEG-2 video standard is introduced, which is measured by average bit-rate saving for different PSNR values that reflect the objective quality of the video frames. Even at High bit-rate, it is clear that the H.264/AVC introduces a valuable saving, when compared to the MPEG-2.

Furthermore, we can draw some conclusions about the H.264/AVC and the MPEG-2 compression efficiency by analyzing Fig. 1.1. First, the H.264/AVC offers a bit-rate gain of roughly fifty percent (50 %) at bit-rate below 15 Mbps. And, at bit-rate above 30 Mbps, the H.264/AVC produces results comparable in quality to MPEG-2 with a 20 Mbps increase. For example, MPEG-2 quality at 60 Mbps is achieved by H.264/AVC at only 40 Mbps or less. As well, at very high bit-rates, this rate saving can sometimes be even greater. However, above the 50 Mbps point, the quality provided by H.264/AVC increases linearly with the rate, showing that most of the encoder effort is spent coding non-redundant information, e.g. noise. Because the human eye is not very sensitive to noise fidelity, most sequences look quasi-transparent above this rate [1].

In addition, H.264/AVC introduces high network adaptation flexibility to error prone networks like mobile channels that have high tendency for bit errors and packet losses. Accordingly, H.264/AVC standard is promised to be the universal video encoder for all network dependent multimedia applications, such as Video on Demand (VoD), video broadcasting and Internet video conferencing, etc.

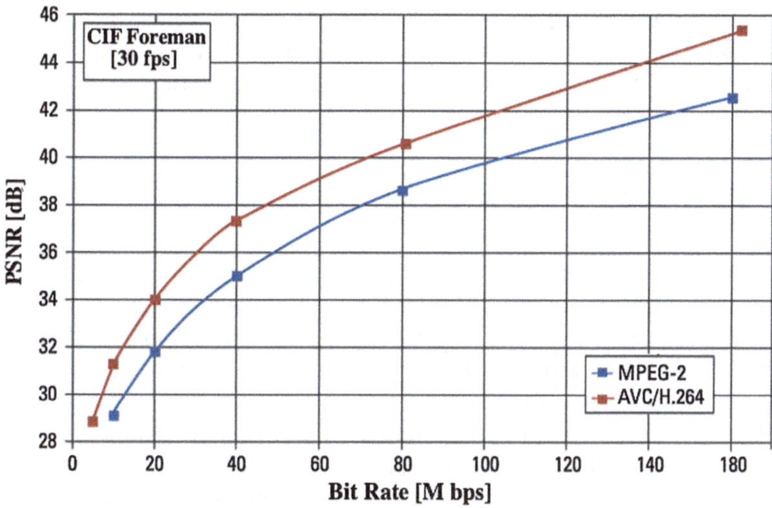

Fig. 1.1 The compression efficiency of H.264/AVC versus the MPEG-2

Moreover, H.264/AVC presents many new coding tools including, but not limited to, variable block size motion estimation, integer DCT, and context-adaptive binary arithmetic coding (CABAC) to achieve better coding performance compared to previous standards [2]. Such enhancement tools enable H.264/AVC to provide high video compression performance for non-interactive applications, like video streaming, which requires high coding efficiency. However, these privileges are achieved at the cost of additional computational complexity, which is caused by the advanced video coding algorithms in the H.264/AVC. The research results have shown that the computational complexity of the H.264/AVC is almost ten times higher than the MPEG-4 Part 2 Advanced Simple Profile (ASP). Therefore, H.264/AVC has not fit efficiently in real-time interactive applications that require low latency systems, e.g. video telephony over mobile devices.

1.1.1 The H.264/AVC Naming Convention

Multiple naming is a common issue in the field of video coding standards. For instance, MPEG-2 is jointly developed by MPEG and the ITU-T, where MPEG-2 is known as H.262 in the ITU-T community.

In the same way, the H.264/AVC standard was developed jointly by Video Coding Experts Group (VCEG) and Moving Picture Experts Group (MPEG), after the initial development work in the ITU-T as a VCEG project, where it was used to be called H.26L. Also, it is common to refer to the H.264/AVC standard with names such as the JVT codec, AVC/H.264, MPEG-2/part10, H.264/MPEG-4 AVC or MPEG-4/H.264 AVC. On one hand, the H.264/AVC name follows the ITU-T naming convention. On the other hand, the MPEG-4 AVC name relates to the naming convention in ISO/IEC

MPEG, where the standard is part 10 of MPEG-4 suite of standards. Moreover, the H.264-compliant decoder must define and be capable of using at least one of the defined sub-set of tools, known as H.264/AVC coding profiles.

1.1.2 The H.264/AVC Coding Profiles

In order to adapt to the different application's quality requirements, the H.264/AVC has three integrated complexity profile. First, "Baseline Profile (BP)", which is the simplest H.264/AVC profile. It introduces the lower complexity bond of the H.264/AVC. For that reason, the Baseline Profile is designed specially to fit time sensitive applications, for instance video on mobile devices and video conferencing.

On the other hand, the most commonly used profile is the "Main Profile (MP)". It targets applications that require high compression rate while preserving the visual quality. Besides, the Main Profile provides higher compression and better quality than the Baseline Profile at the cost of significant computational complexity. In particular, the Digital TV (DTV) broadcasting is an example for video application that uses the H.264/AVC Main Profile. The third profile, which is known as the "Extended Profile (EP)", is optimized for efficient video streaming applications. EP has relatively high compression capability. Additionally, it offers some extra tricks to overcome data losses and server stream switching.

Recently, another two profiles have been developed the Multi-view Video Coding (MVC) extension and the Stereo High Profile (SHP), which have been finalized on November 2009. The MVC High Profile supports an arbitrary number of views. On the other hand, the SHP is specifically designed for two-view stereoscopic video.

In short, a specific decoder may decode one or more profiles, but it is not necessarily capable to decode all the H.264/AVC profiles. Also, the decoder specification should describe which of the profiles can be decoded. In the scope of this book, we are mainly targeting the video mobile applications and video conferencing. Accordingly, we have adapted the Baseline Profile in all of the introduced studies and experiments.

1.1.3 The Scope of the H.264/AVC Standard

The standard document does not actually specify how to encode the source video, which is left to the manufacturer of the video encoder. However, in practice, the encoder is likely to mirror the steps of the decoding process. As illustrated in Fig. 1.2, the structure of the H.264/AVC encoder is a mirror for the standard H.264/AVC decoder. However, the video output is not identical to the video source because H.264/AVC is a lossy compression standard. And, some of the video subjective and objective qualities are lost during the compression and the decompression processes.

Furthermore, during the H.264/AVC coding processes, the raw video data is processed in block units called macro-block, which is corresponding to 16×16

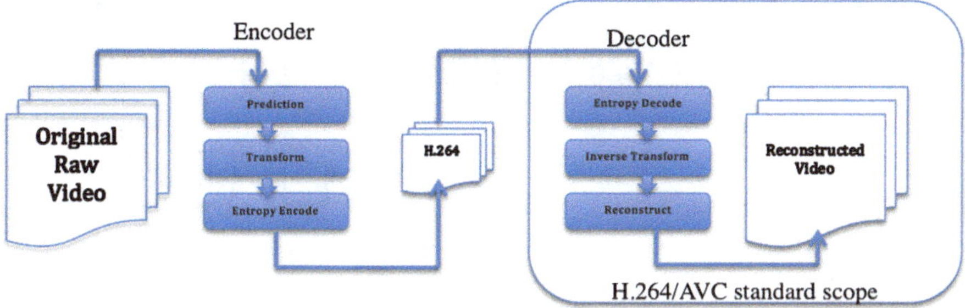

Fig. 1.2 The scope of the H.264/AVC video standard

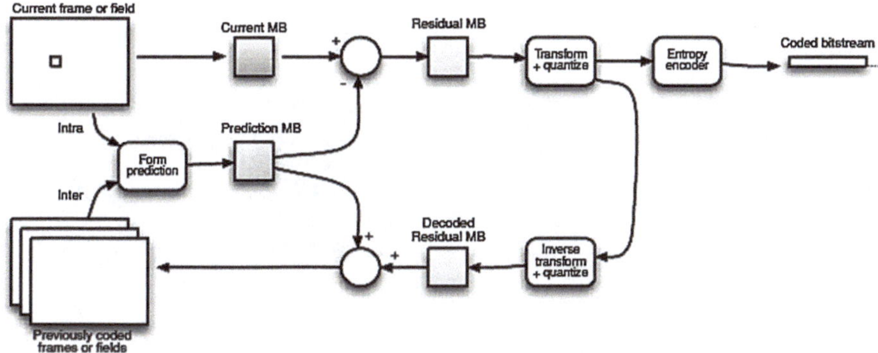

Fig. 1.3 The architecture of H.264/AVC standard encoder

pixels of the video frame. In the H.264/AVC encoder, a predicted macro-block is generated and subtracted from the current macro-block to form a residual macro-block, which is transformed, quantized and then encoded. In parallel, the quantized data is re-scaled and inverse transformed, then added to the predicted macro-block to reconstruct a coded version of the macro-block, which is stored for later predictions.

On the other side, in the H.264/AVC decoder, the coded macro-block is decoded, re-scaled and inverse transformed to form a decoded residual macro-block. After that, the decoder generates the same prediction that was created at the encoder, then adds it to the residual to produce a decoded macro-block. Figures 1.3 and 1.4 illustrate the standard H.264/AVC standard encoder architecture blocks and the corresponding standard H.264/AVC decoder building blocks, respectively [7].

1.1.4 The H.264/AVC Prediction Based Compression Technique

The H.264/AVC video standard predicts the current macro-block based on a previously encoded macro-block. Such previously encoded data is, either from

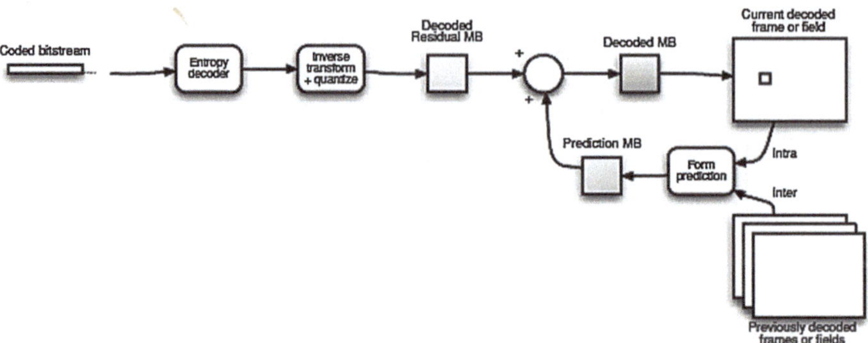

Fig. 1.4 The architecture of H.264/AVC standard decoder

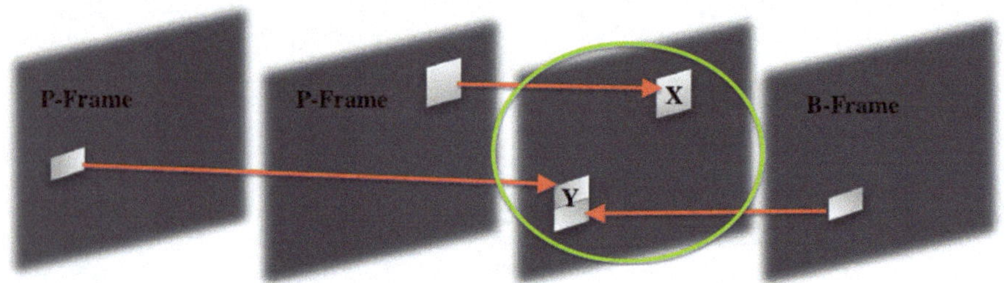

Fig. 1.5 The H.264/AVC standard inter prediction technique

surrounding and previously coded pixels within the current frame (I-Frame) in case of using the Intra prediction. Or, from other previously encoded frames (P-frame) and/or future frames (B-frame), in case of using the Inter prediction. Then, it subtracts the predicted macro-block from the current macro-block to form the residual. Furthermore, the Inter prediction process uses a range of block sizes from 16×16 down to 4×4 to predict pixels in the current frame from similar regions in previously coded frames. These previously coded frames may occur before, which is called P-frames or, after the current frame in the display order, which is called B-frames.

In Fig. 1.5, the macro-block "X" is the current macro-block, which is predicted from a 16×16 macro-block in the previous encoded P-Frame. In contrast, the macro-block "Y" is predicted from two different frames. The upper 8×16 sub macro-block is predicted from a P-Frame, which is two frames away from current frame. While the lower 8×16 sub macro-block is predicted from a B-frame, which is the next coming frame in the timely order.

Figure 1.6 shows an example of 16×16 macro-block and its prediction from a previous and future frames using the same prediction scenario as in Fig. 1.5. In this example, the Inter predicted 16×16 macro-block, in the middle figure, is a good match for the current macro-block. Accordingly, the values in the residual macro-block, as shown in the left figure, are zeros or mostly near zero. In the scope of this

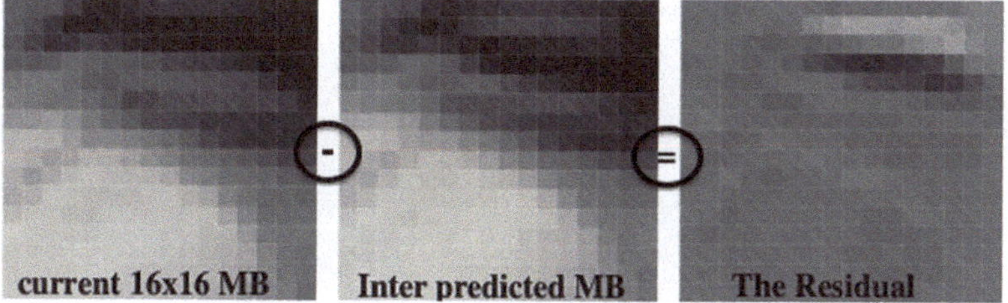

Fig. 1.6 Example for inter prediction on 16 × 16 macro-block

Fig. 1.7 Intra prediction modelling using roster scan order

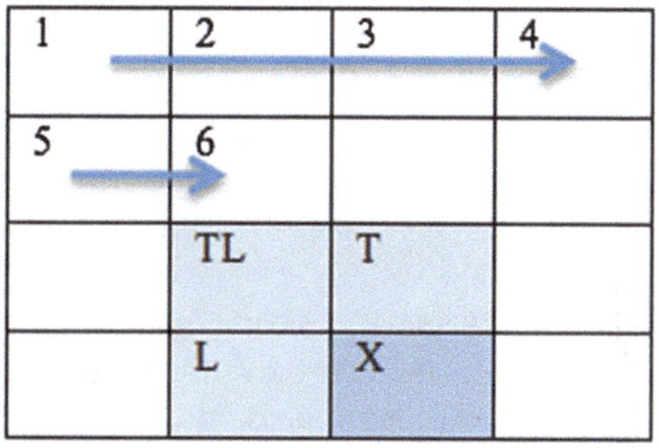

book, we are more interested in the Intra prediction, which the Sect. 1.1.5 discussing with slightly more detail, in the standard H.264/AVC baseline profile.

1.1.5 The Intra Prediction in H.264/AVC Standard Baseline Profile

The performance of H.264/AVC Intra frame encoding is depending on its computation intensive coding tools such as multiple modes Intra prediction, Lagrangian-based mode selection and entropy coding algorithms [3]. Indeed, the Intra prediction is a very critical process in H.264/AVC encoder as the image's spatial correlation is used to dramatically improve the compression ratio. Also, it plays an important rule in rising up the encoder's overall performance.

In the Intra prediction process, every macro-block is encoded by using previous encoded neighboring macro-blocks' samples within the same frame. As shown in Fig. 1.7, the Intra prediction process can be modelled, at the macro-block level, in the encoder side as in (1.1) and (1.2). Similarly, (1.3) and (1.4) are modelling the Intra operation at the decoder side of the H.264/AVC codec. In the H.264/AVC Encoder:

$$P(X) = (2L' + T' + TL')/4 \tag{1.1}$$

$$R(X) = X - P(X) \tag{1.2}$$

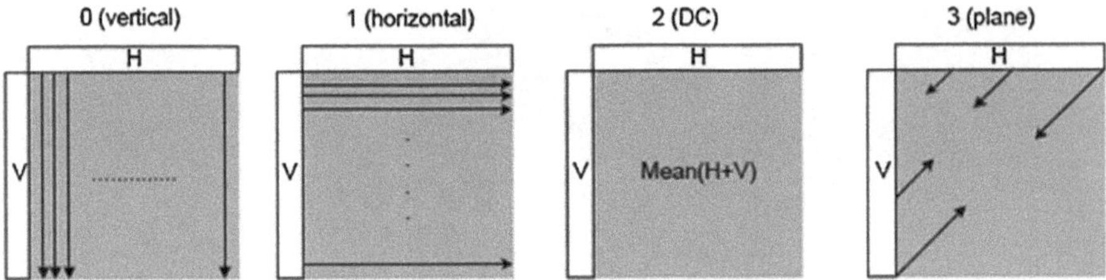

Fig. 1.8 The reconstruction directions for the intra 16 × 16 mode

In the H.264/AVC Decoder:

$$P(X) = (2L' + T' + TL')/4 \tag{1.3}$$

$$X = R(X) + P(X) \tag{1.4}$$

where L′, T′ and LT′ are the reconstructed version of macro-block L, T and LT, respectively. And, X is the currently processed macro-block.

Clearly, the Intra prediction coding performance is comparable to that of JPEG2000 [4]. Therefore, the Intra frame encoding that is currently used in H.264/AVC, is suitable for both image and video compression. However, the high computational complexity of Intra prediction process has been a barrier for employing H.264/AVC to interactive real-time video applications.

1.1.6 The H.264/AVC Intra Prediction Modes in Baseline Profile

In H.264/AVC baseline profile, the Intra prediction supports two Intra prediction modes for the macro-block Luma component, which are the Intra 4 × 4 prediction mode and the Intra 16 × 16 prediction mode. However, the Chroma components have only one single mode, which is the Intra 8 × 8 prediction mode.

In the Luma 16 × 16 macro-block mode, each 16 × 16 macro-block can select one of four reconstruction directions, which are illustrated by Fig. 1.8. Also, each direction has different methodologies to reconstruct the current 16 × 16 macro-block. For instance, mode (3) is called plane prediction, which is approximated bilinear transform using only integer arithmetic.

On the other hand, in case of the Luma 4 × 4 macro-block mode, each of the 4 × 4 macro-blocks can select one of nine prediction modes, which are shown in Fig. 1.9. In additional to the DC prediction, eight directional prediction modes are used for reconstructing the 4 × 4 macro-block. Also, the prediction process utilizes up to 13

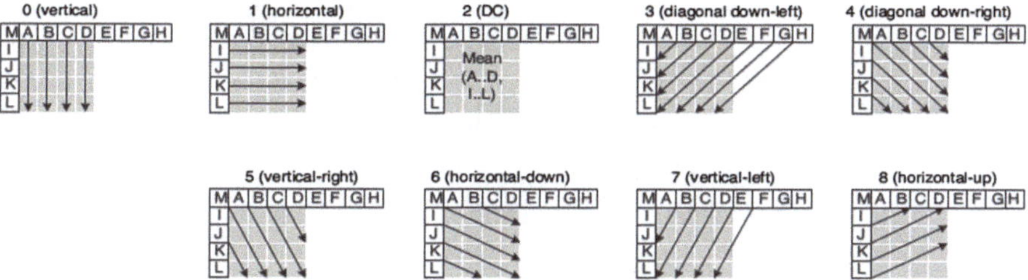

Fig. 1.9 The reconstruction directions for the Intra 4×4 mode

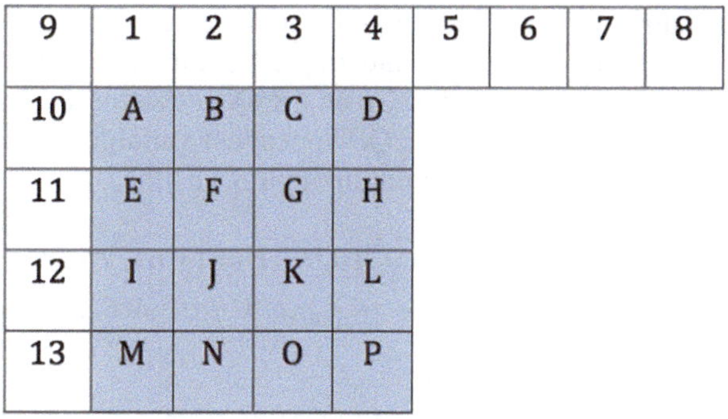

9	1	2	3	4	5	6	7	8
10	A	B	C	D				
11	E	F	G	H				
12	I	J	K	L				
13	M	N	O	P				

Fig. 1.10 The neighboring pixels used by the Intra 4×4 mode

directly neighboring pixels from previously coded and reconstructed macro-blocks as illustrated in Fig. 1.10.

Furthermore, the Intra predictions for each of the Chroma C_r and C_b components are completely independent of the Luma component prediction. Also, both Chroma components are simultaneously predicted in the baseline profile at one mode only, which is the 8×8 macro-block mode. Moreover, the Chroma prediction modes are very similar to those of the Luma 16×16 macro-block mode. The only difference, beside the different block size (8×8), is a very fine modification in calculating the "DC" prediction mode.

Obviously, the bit-rate results from using the Luma 16×16 mode is less than the bit-rate of the 4×4 macro-block mode. Also, the analytical results to huge sets of standard video frames benchmarks confirmed that the 4×4 macro-block mode tends to be used for highly textured regions. In contrast, the 16×16 macro-block mode is more likely to be chosen for smooth regions.

Although the mode decision process is the most important step, it has not been specified in H.264/AVC standard. It has been left as an encoder issue due to its effects on the overall coding performance. However, the reference software of H.264/AVC recommends two different mode decision methods, i.e. the low complex-

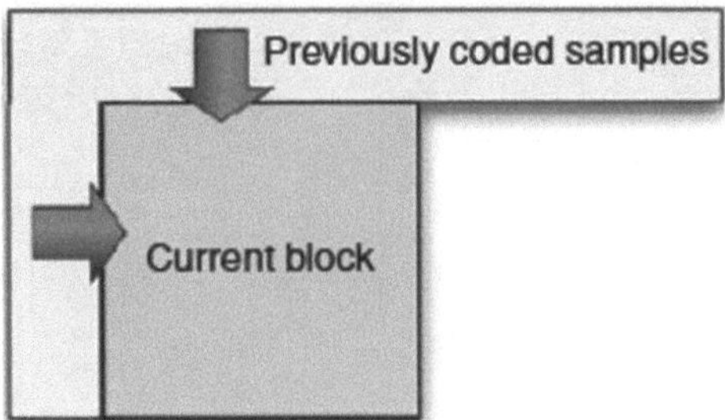

Fig. 1.11 The H.264/AVC intra spatial extrapolation

ity and the high complexity mode decision. Both mode decision methods choose the best possible macro-block mode by evaluating a Lagrangian Cost Function (LCF) that considers both distortion and rate. Given the Quantization Parameter (QP) and the Lagrange Parameter (λ), which is a QP-dependent variable, the mode decisions for each macro-block are derived by minimizing Eq. (1.5).

$$J_{mode}(MB_k, Ik|QP, \lambda_{mode}) = Distortion(MB_k, I_k|QP) + \lambda_{mode} * Rate(MB_k, I_k|QPP) \tag{1.5}$$

Accordingly, for the low complexity mode decision, the distortion is evaluated by using the Sum of Absolute Differences (SAD) between the predicted and original versions of the pixels. Even though, the coding performance can be enhanced by 0.2–0.5 dB better, if the Sum of Absolute Transformed Differences (SATD) are considered instead of the SAD. However, only the number of bits required to code the mode information estimates the rate.

On the contrary, in case of the high complexity mode decision is considered, the distortion is calculated by using the Sum of Square Differences (SSD) between the raw samples and the reconstructed pixel values. However, both the number of bits required for coding the mode information and the residual estimate the rate. Unfortunately, the high complexity mode introduces an unacceptably high computational complexity, especially for real-time applications. Regardless the complexity, it leads to the best compression performance possible.

1.1.7 H.264/AVC Intra Spatial Extrapolation

The spatial extrapolation is used by the H.264/AVC to create a precise macro-block Intra prediction. In other words, each macro-block within the current I-frame is predicted from previously coded macro-blocks, which are also within the same Intra frame (I-frame). As shown in Fig. 1.11 [7], in case of using the roster-scan order to process the current I-frame, all the upper macro-block's rows and all the macro-blocks

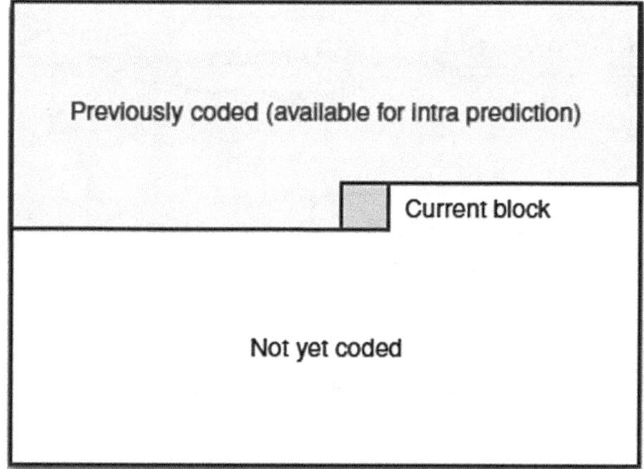

Fig. 1.12 The available samples for H.264/AVC intra prediction

to the left of the current macro-block are available to be used by the H.264/AVC Intra prediction process. Also, the encoded bit stream representing these macro-blocks have been already decoded and ready to be used in the recreation process of the current macro-block. During the Intra prediction process, the predictions are created by extrapolating samples from top and/or left sides of current macro-block as shown in Fig. 1.12 [7].

After the predicted macro-block has been reconstructed, the residual is calculated by subtracting the current macro-block from the predicted macro-block. Then, the residual is transformed and encoded along with the prediction mode and the reconstruction direction. Obviously, if the prediction is successful, the energy in the macro-block's residual is lower than the energy in the original macro-block. As a result, the residual can be represented with fewer bits.

1.1.8 The H.264/AVC Integer DCT and Quantization

In the H.264/AVC codec, the output of Intra prediction stage, at the macro-block level, is a macro-block size of residual samples, which are eventually the input for the transformation stage. Furthermore, the values of the residual macro-block are transformed using a 4×4 or 8×8 integer approximation of the Discrete Cosine Transform (iDCT). On the other hand, the output of the transform stage is a set of DCT coefficients, each of which is a weighted value for a standard basis pattern.

Next, each of the iDCT coefficient is quantized that is dividing the coefficient value by an integer quantization parameter (QP). Such process reduces the precision of the coefficient's values by the value of the QP. Furthermore, the resulting values from the division process are rounded to the nearest integer. In fact, most of the quantized macro-block coefficients are zeros. However, the number of zero and non-zero

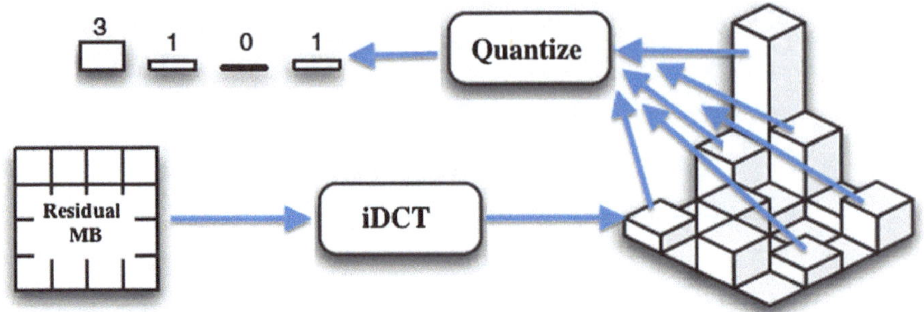

Fig. 1.13 iDCT transform and quantization in H.264/AVC encoder

coefficient in the macro-block is depending on the QP value. In other words, the higher the QP the more coefficients are set to zero. As a result, a higher compression is achieved at the expense of poor decoded image quality. Accordingly, by setting the QP to a low value, more non-zero coefficients remain after the quantization, which results in a better image quality at the cost of lower compression rate. Figure 1.13 illustrates both the iDCT and the quantization process in the H.264/AVC encoder side.

On the other hand, in the H.264/AVC decoder side, the quantized transform coefficients are first re-scaled by multiplying each coefficient by the same integer QP value used by the encoder. Such process is well known as inverse quantization (q^{-1}). However, the re-scaled transformed output values are not identical to the corresponding transformed coefficient in the encoder side as the quantization process is not fully reversible.

Moreover, as shown in Fig. 1.14, the rescaled coefficients are inverse transformed to recreate the macro-block residual values. Such process is sometimes known as inverse Discrete Cosine Transform (DCT^{-1}). The output of the DCT^{-1} stage is the residual macro-block samples, which are not identical to the one in the encoder side. In fact, the forward quantization process causes the difference, which is sometimes known as "energy loss". Obviously, a larger QP tends to produce a larger difference or more energy loss between the original and the reconstructed macro-blocks.

1.1.9 The H.264/AVC Entropy Encoding and Reconstructing

The H.264/AVC video coding processes convert the raw video samples into a compressed bit-stream, which includes encoded values that represent information about the over all video sequence, information about the compression parameters, information about the prediction modes and directions and information about the quantized transformed residuals. Then, the Variable Length Coding (VLC) and the

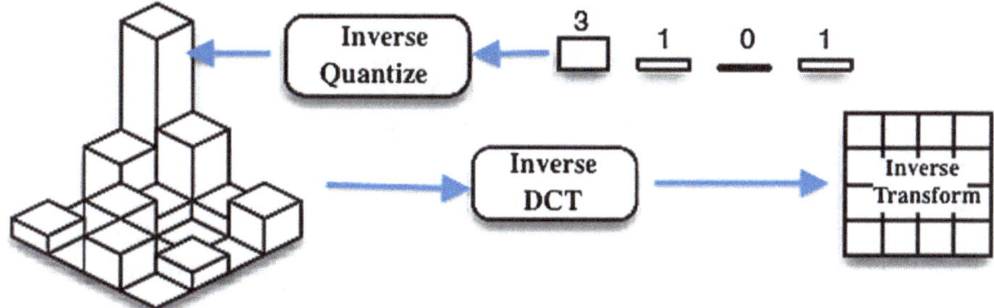

Fig. 1.14 Rescaling and inverse DCT transform in H.264/AVC decoder

Arithmetic Coding (AC) are used to convert all the encoded values to a binary code, which is a binary representation of the information. Finally, the produced H.264/AVC compressed bit-stream is ready to be stored or transmitted over the network, e.g. the Internet.

On the other hand, the H.264/AVC decoder initiates the decoding process by extracting the above information from the compressed H.264/AVC bit-stream, i.e. the quantized coefficients of the macro-block residuals, the quantization parameter and the prediction information. Finally, this information is used to reverse all the encoding process and reconstruct the video frames.

1.2 Problem Statement

The MPEG-2 video compression standard is widely deployed not only in video distribution infrastructures like Digital Television (DTV) cable and satellite, but also in several popular applications such as DVDs and DVRs. Additionally, for many years, end-to-end systems have existed with millions of interoperable MPEG-2 encoders and multiplexers deployed. However, MPEG-2 bit stream requires almost double the transmission bit rate and larger storage capacity compared to what H.264/AVC consumes for the same picture quality [5].

Moreover, the H.264/AVC has been a break through in the field of video coding. It has been widely implemented in most of the mobile multimedia terminals because of its high compression rate while the quality degradation cost is preserved to the minimum [7]. In addition, for its high network adaption capability, the H.264/AVC video standard is promised to be the universal video codec for all network dependent multimedia applications, i.e. Internet video applications, mobile video applications and video conferencing, etc.

Consequently, the need for MPEG-2 to H.264/AVC transcoder has rapidly grown, not only because of the fast spread of multimedia enable mobile devices, but also the large number of MPEG-2 video contents that already exist. In order to access such MPEG-2 contents from devices with different network capabilities, the need for an efficient transcoder has been arisen. At the same time, the high industrial demand on such transcoder presents many research challenges. High throughput and power efficient are the most demanded features of such transcoders.

Furthermore, the Conventional Cascaded Transcoder (CCT) is a direct well known, yet naive, MPEG-2 to H.264/AVC transcoder. It is achieved by cascading MPEG-2 decoder and H.264/AVC encoder. Although it represents an upper limit on computational complexity, it theoretically maximizes the video quality [5]. In particular, the main problem in transcoding came from the differences between H.264/AVC and MPEG-2 [6]. However, a computationally efficient transcoder would take advantage of the information stored in MPEG-2 decoder to accelerate the recoding process in the H.264/AVC encoder [14].

Additionally, the research results have shown that the computational complexity of the H.264/AVC is almost ten times higher than the MPEG-4 Part 2 Advanced Simple Profile (ASP). Therefore, H.264/AVC has not fit efficiently in real-time interactive applications specially that require low latency and low power systems, e.g. video telephony over mobile devices. In particular, the high computational complexity of Intra Prediction process has been one of the main barrier for employing H.264/AVC to interactive real-time video applications.

This book explores the MPEG-2 to H.264/AVC from the industrial point of view. It defines a critical cause for the huge computational complexity of the subjective transcoder, specifically the Intra mode and direction prediction of the H.264/AVC part of the transcoder. This book's main objective is to design and optimize a low complicity Intra mode and direction algorithm to overcome the computational complexity issue in the industrial H.264/AVC to MPEG-2 transcoder. Furthermore, Iwe targeted to standardize our proposed Intra mode and direction prediction solution by isolating and framing it to suite into the standard H.264/AVC JM standard software. In addition, this book introduces a low-power hardware architecture design for the proposed high-throughput Intra prediction technique. The hardware solution has been optimized for video application in mobile devices.

1.3 Contributions and Results

In this book, there are several distinct contributions. Here, they are listed in the same order they appear:

- A brief overview over the traditional industrial video transcoding techniques (Chap. 2).
- A survey of related literature in the area of Intra prediction compression in the H.264/AVC video standard (Chap. 3).
- Our MPEG-2 to H.264/AVC real-time transcoding scheme (Chap. 4).
- Our simulation, comparison and analysis showing the benefits of the proposed MPEG-2 to H.264/AVC transcoding scheme (Chap. 4).
- Our enhanced Intra prediction algorithm entitled "Full-Search Free (FSF) Intra Prediction Algorithm" (Chap. 5).
- Our simulation, comparison and analysis showing the advantages of the proposed FSF Intra prediction scheme (Chap. 5).

- Our hardware architecture design and its FPGA and ASIC implementations for our proposed FSF Intra prediction scheme (Chap. 5).

Indeed, the FSF algorithm and the real-time transcoding scheme are new ideas that do not exist in the current literature. At first, the FSF in Chap. 5 is a new algorithm that has been developed to enhance the computational complexity of the Intra prediction process in the H.264/AVC video standard. Moreover, the FSF main enhancement is in its ability to reduce the run-time while maintaining the same visual quality as the standard H.264/AVC Intra prediction algorithm. Also, the reduction in the Intra prediction time qualifies the H.264/AVC codec for real-time applications. Furthermore, the FSF hardware architecture, also introduced in Chap. 5, is a new Intra prediction design that granted the required operating frequency for real-time H.264/AVC codec devices.

Moreover, another new contribution is the MPEG-2 to H.264/AVC transcoding algorithm in Chap. 4. The achieved enhancement is in the algorithm ability to significantly reduce the transcoding time while maintaining the same visual quality as a transcoding system without our algorithm. Also, the reduction in the transcoding time nominates the proposed enhanced transcoder for real-time applications.

On the other hand, the rest of the listed contributions are introduction and literature survey type of contribution. All in all, the main contribution in this book is bi-folded. On one hand, it advances the work in the field of heterogeneous video transcoding. On the other hand, it qualifies the H.264/AVC codec for real-time video applications and devices.

Chapter 2
Conventional Transcoder

Abstract The H.264/AVC video compression standard offers over two times higher compression rate than what the MPEG-2 can offer, while maintaining the same visual quality. Additionally, H.264/AVC offers a substantial network adaptation capability, which is essential for many network dependent video applications, e.g. delivering Digital TV (DTV) programs over the Internet and video conferencing. For that reason, the relativity new H.264/AVC standard has been pushed to the top of the candidate stack for coding Digital TV programs at a bit-rate lower than the MPEG-2's bit-rate. However, for over a decade now, most of the video resources have been encoded in MPEG-2 format. As a result, the heterogeneous transcoding solution, MPEG-2 to H.264/AVC, becomes necessary for distributing video resources specially over the Internet. In the following section, we will discuss the current employed MPEG-2 to H.264/AVC transcoding solution in the video industry.

2.1 The MPEG-2 to H.264/AVC Conventional Cascaded Transcoding

The simplest transcoding method is the one that converts the MPEG-2 video formats to the H.264/AVC bit-stream by means of cascaded transcoding. Also, it is well known in the video industry with its commercial name, i.e. "Conventional Cascaded Transcoder (CCT)". Simply stated, the idea behind such transcoder is based on sequential decoding/ encoding processes. At first, the MPEG-2 video file is completely decoded using a traditional MPEG-2 decoder. After that, the uncompressed video frames, which also entitled either "Raw Video", uncompressed video or Common Intermediate Format (CIF), are encoded by an independent H.264/AVC encoder to produce the H.264/AVC bit-stream.

As illustrated in Fig. 2.1, the MPEG-2 bit-stream is decoded by the MPEG-2 decoder to produce the reconstructed raw video data. Then, the uncompressed video frames are compressed by the H.264/AVC encoder to generate the H.264/AVC compressed video bit-stream.

T. Elarabi et al., *Real-Time Heterogeneous Video Transcoding for Low-Power Applications*, DOI: 10.1007/978-3-319-06071-2_2,

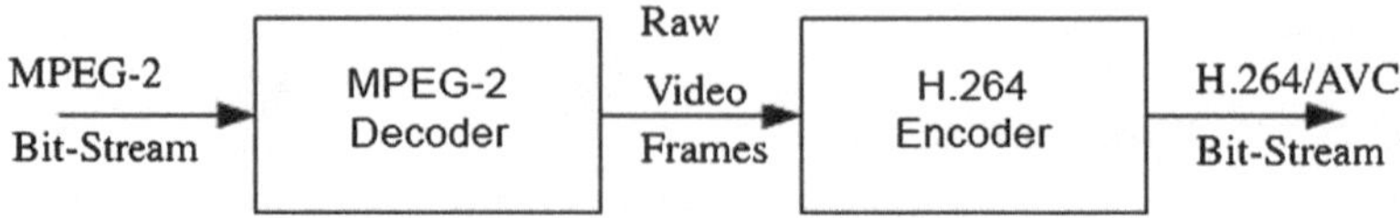

Fig. 2.1 The conventional cascaded MPEG-2 to H.264 transcoder

In fact, in terms of hardware and software implementation, the cascaded transcoding architecture is implementation friendly. Also, it can reduce the lab-to-market time gap, besides, the engineering cost [8]. Furthermore, it has been widely used in the market because of its conceptually straight forward design structure and relativity easily implementation. However, the cumulated computational complexity is causing a serious problem for the cascaded transcoding solution.

Indeed, the computational complexity of the H.264/AVC decoder is ten times or more than the computational complexity of the MPEG-2 encoder. Accordingly, most of the computational complexity is localized in the H.264/AVC encoder side of the conventional transcoder. In contrast to the MPEG-2 codec, the H.264/AVC standard codec permits not only Inter frame coding to remove the frames temporal redundancy, but also Intra frame coding to eliminate the frames' spatial redundancy. In particular, both the Intra macro-block mode decision and the Intra macro-bock direction prediction processes are the main computational demanding stages in the Intra frame coding process [9].

Furthermore, the H.264/AVC has two prediction mode candidates for the macro-block's Luminance (Luma) component. On the other hand, it has only one mode for each of the macro-block's Chrominance (Chroma) components. Accordingly, the mode selection process is done only on the macro-block's Luma component. Briefly, the two Luma Intra prediction modes are the Intra 16 $\times$ 16 mode or the Intra 4 $\times$ 4 mode in the H.264/AVC baseline profile [7].

On one hand, the Intra Luma 16 $\times$ 16 mode executes the direction prediction process on the whole 16 $\times$ 16 macro-block as one unit. In contrast, the Intra Luma 4 $\times$ 4 mode performs the direction prediction process on each of the sixteen 4 $\times$ 4 sub macro-blocks within the current 16 $\times$ 16 Luma macro-block. Furthermore, there are four reconstruction direction candidates for the Intra Luma 16 $\times$ 16 mode, as shown in Fig. 2.2. Whereas, there are nine direction prediction candidates for the Intra Luma 4 $\times$ 4 mode, as shown in Fig. 2.3.

On the other hand, there is only one Intra 8 $\times$ 8 prediction mode for the Chroma macro-block components [10]. It has four different direction prediction candidates, which are similar to Intra 16 $\times$ 16 Luma direction prediction candidates except that the numbers of the modes are using different order as shown in Table 2.1 [7].

Furthermore, in the conventional cascaded pixel domain transcoder, the Rate Distortion (RD) is used to evaluate and further decide the reconstruction direction for all of the macro-block's components. In details, the RD for all the four Intra Luma 16 $\times$ 16 direction prediction candidates, all the nine Intra Luma 4 $\times$ 4 direction prediction candidates and the four Chroma direction prediction candidates for

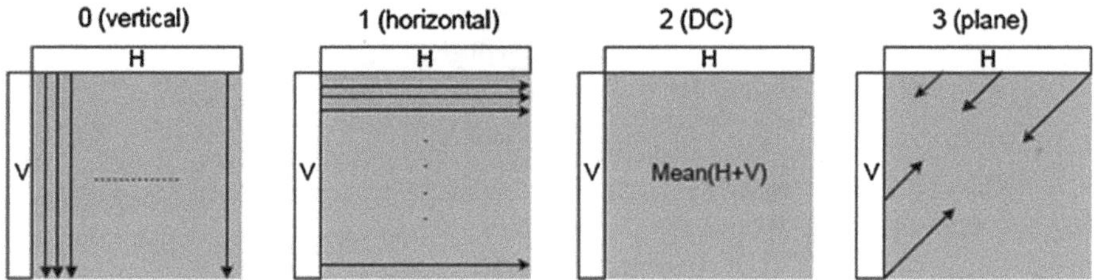

Fig. 2.2 The standard direction candidates of the 16 × 16 mode

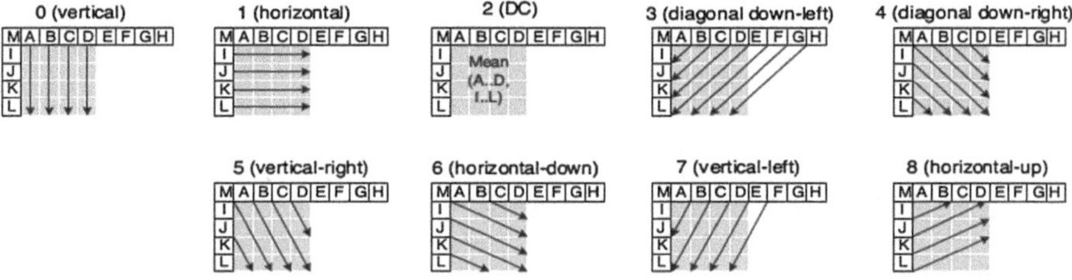

Fig. 2.3 The standard direction candidates of the 4 × 4 mode

Table 2.1 The Luma versus the Chroma direction candidates' number

Prediction direction	Luma number	Chroma number
DC	2	0
Horizontal (H)	1	1
Vertical (V)	0	2
Plane (P)	3	3

each macro-block must be computed. In other words, in order to make a mode and reconstruction direction decision for each of the macro-block's components, the cost is evaluated for all the macro-block direction prediction candidates based on the RD theory [11] as in Eqs. (2.1) and (2.2).

$$RD_{cost} = D + LM * BR \tag{2.1}$$

$$LM = 0.85 * 2^{(QP-12)/3} \tag{2.2}$$

where D and BR represent the distortion and bit-rate for a given prediction direction, respectively. QP and LM are the quantization parameter and Lagrangian multiplier, respectively. Also, in order to compute BR and D, the actual coding is performed for each 4 × 4 sub macro-block, which is the basic processing unit for transformation and Variable Length Coding (VLC) in the H.264/AVC codecs.

All in all, as shown in Fig. 2.4, the total number of needed Rate Distortion Operations (RDO) for deciding the best direction for each macro-block is 592 RDOs

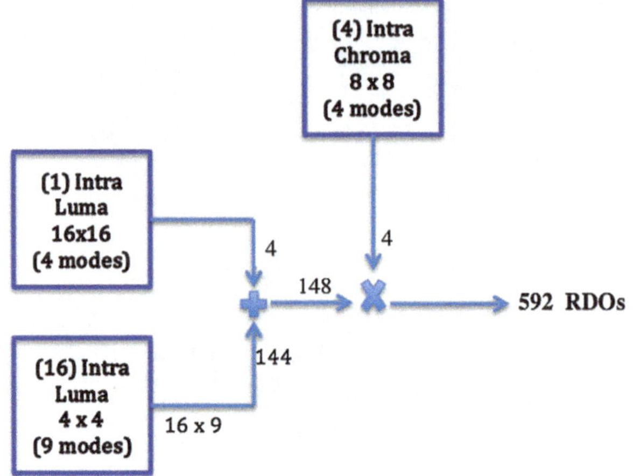

Fig. 2.4 The required RDOs for each standard Intra macro-block

[9]. Such huge computational complexity is the current bottleneck in the MPEG-2 to H.264/AVC conventional cascaded transcoder processes.

In conclusion, the computational complexity of the H.264 encoder part of the conventional cascaded transcoder is proved to be impractically huge. Besides, it requires multiple Digital Signal Processing (DSP) cores, which demand sophisticated parallel processing algorithms [8]. Moreover, this implementation issues are very serious for real-time video communication, e.g. real-time TV broadcasting to H.264/AVC enabled smart phones and video telephony [8].

Additionally, the H.264/AVC video decoders in the mobile devices periodically use the Intra frame prediction to quickly recover from channel error propagation at rate of one Intra frame per every five frames [9]. Whereas, the Intra prediction built in processes is the cause for most of the H.264/AVC computational complications. Therefore, the main scope of the thesis is focusing on enhancing the H.264/AVC Intra prediction side of the transcoder. However, there are many research efforts that have been introduced to reduce the computational complexity of the conventional transcoders. Also, many researchers have handled the Intra prediction complexity issue at the H.264/AVC encoder part of the CCT transcoder. As a result of such research efforts, several enhanced transcoding algorithms have been recently introduced.

For example, a very interesting work was introduced in [8]. In this research work, the authors proposed a mode skipping rule for the H.264/AVC encoder's Intra prediction process in the MPEG-2 to H.264/AVC transcoder. Their idea is based on an experimental analysis results, which show that the DCT energy trend in the MPEG-2 bit-stream has a strong correlation with the Intra mode selection in the H.264/AVC encoder. In Chap. 3, we discuss their Intra prediction technique, which is entitled "Kim's algorithm" in some details. Additionally, in Chap. 5, we compared

Table 2.2 Brief comparison between MPEG-2 and H.264/AVC

	MPEG-2	H.264/AVC
Transformation	DCT	DCT
Inter prediction	Yes	Yes
Intra prediction	No	Yes
MB partition	No	Yes
Block sizes	$16 \times 16, 8$	$16 \times 16, 8$; $8 \times 16, 8, 4$; $4 \times 8, 4$
Reference frames	1 or 2	Up to 16

Kim's Intra prediction algorithm with our proposed Full-Search Free Intra prediction algorithm, which we introduced in Sect. 5.2 of the same chapter.

In addition, the authors of [9] proposed an Intra mode decision method for the MPEG-2 to H.264/AVC transcoder, which is based on a spatial activity analysis for the DCT coefficients in the MPEG-2 decoder part of the transcoder. Chapter 3, discusses Yoo's Intra prediction algorithm in details. Also, Chap. 5 includes a comparison between Yoo's Intra algorithm and our proposed Full-Search Free Intra prediction algorithm. Furthermore, analyzing the similarities and the differences between both the H.264/AVC and MPEG-2 video compression standards is a good start in order to design an efficient transcoder.

2.2 The MPEG-2 versus H.264/AVC Video Standards

In fact, many transcoding challenges have been arisen due to the differences between the MPEG-2 and the H.264/AVC standard formates. These differences can be summarized, but not limited to the fact that the MPEG-2 standard does not use any Intra compression technique to remove the Intra frames' spatial redundancy. However, the H.264/AVC standard encoder uses the Intra prediction technique quite frequently for compressing the reference frames (I-frames). In brief, Table 2.2 summarizes the main differences between the MPEG-2 and the H.264/AVC standards.

In fact, the H.264/AVC's Inter prediction uses different block sizes than the MPEG-2's motion estimation's block sizes. As the MPEG-2 standard is using 16×16 and 8×8 macro-block sizes, the H.264/AVC has a larger set of block sizes: 16×16, 16×8, 8×16, 8×8, 8×4, 4×8, and 4×4. Also, MPEG-2 uses only 1/2-pixel accuracy for motion estimation and compensation. In contrast, H.264/AVC is using up to 1/4-pixel. Similarly, the motion vectors used in MPEG-2 are not constrained to the current frame boundaries. But, it uses only one reference frame for P-type frames that is a previously encoded frame and two reference frames for B-type frames, which are future frames [12]. On the other hand, the H.264/AVC may reference up to eight P-Frames and/or eight B-Frames.

All in all, there are many critical differences between the MPEG-2 and the H.264/AVC video compression standards. As a result, many challenges are facing the industrial demand for an efficient real-time MPEG-2 to H.264/AVC transcoders. In the scope of this book, the Intra prediction and the block sizes are the most concerning differences because of their dramatically high run-time and computational resources consumption. Also, we are focusing on the DCT transformation in both standards as the most concerning similarity between standards from our research point of view.

Chapter 3
Efficient MPEG-2 to H.264/AVC Transcoding

Abstract There are several research efforts that have been introduced over the past decade to address the computational complexity in the MPEG-2 to H.264/AVC transcoder. This chapter introduces a brief literature survey of related and inspiring work to address such computational complexity problem.

3.1 Introduction

The work in [13] is a good example that address the computational complexity issue in the MPEG-2 to H.264/AVC transcoder. The authors used the 8 × 8 DCT coefficients from MPEG-2 decoder side of the transcoder to propose a low complexity Intra mode selection algorithm. Wang's algorithm reduces the computational complexity by less than half (55.44 % CPU time reduction) compared to the conventional transcoder complexity. However, as shown in Figs. 3.1 and 3.2, this computational complexity reduction came with unacceptable cost of more than 25 % increase in the bit-rate of a given PSNR. Such compression efficiency loss takes away the main privilege from using the H.264/AVC, which is the lower bit-rate compressed video files at the same visual quality.

Furthermore, in [14], the authors propose a transform-based solution for the computational complexity problem in the MPEG-2 to H.264/AVC transcoder. Kalva's algorithm splits and recombines the DCT coefficient of the MPEG-2 8 × 8 macro-blocks to 16 × 16 and 4 × 4 macro-blocks in order to adapt to the block sizes of the H.264/AVC standard. Additionally, their introduced algorithm uses the macro-block mean and variance to skim some of the Intra direction candidates. Also, Kalva's approach completely avoids some of the mode's calculations and direction prediction's computations in the H.264 encoding stage in order to reduce some of the transcoder computational complexity. As illustrated by Figs. 3.3 and 3.4, Karva's transcoding algorithm was able to overcome Wang's drawback by maintaining almost the same bit-rate of a given PSNR as the H.264/AVC standard.

T. Elarabi et al., *Real-Time Heterogeneous Video Transcoding for Low-Power Applications*, DOI: 10.1007/978-3-319-06071-2_3,

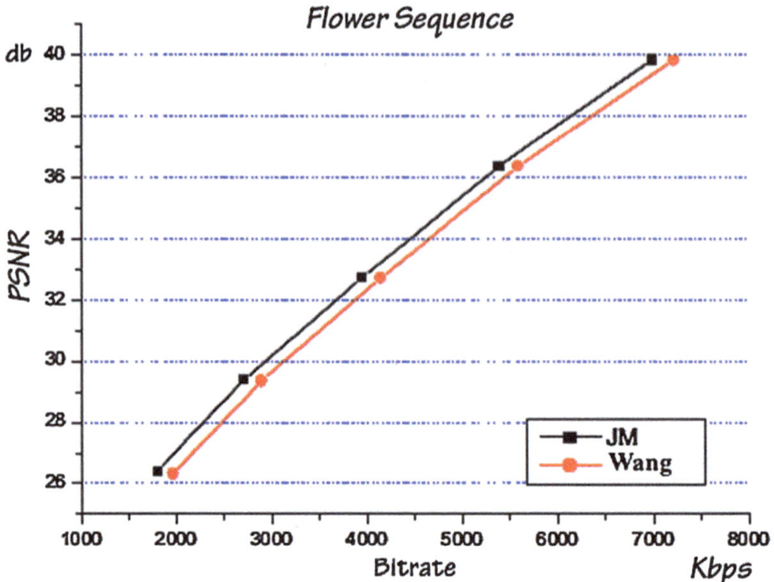

Fig. 3.1 Wang's transcoder RD performance on *Flower sequence*

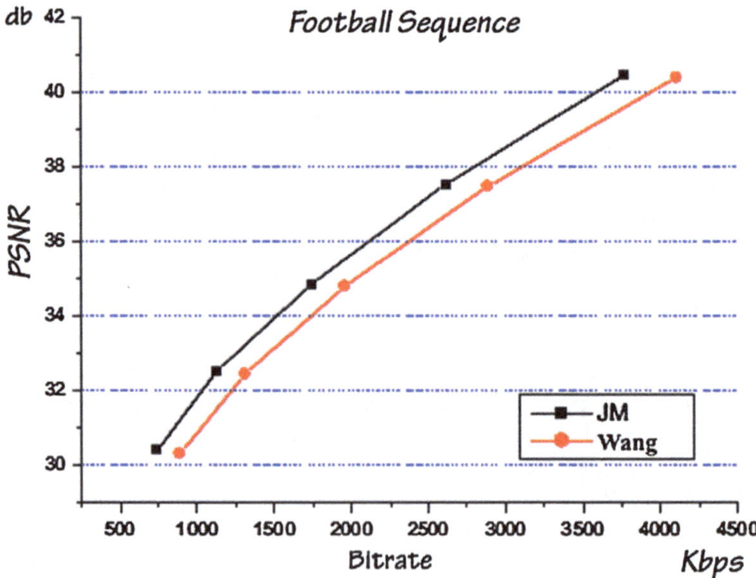

Fig. 3.2 Wang's transcoder RD performance on *Football sequence*

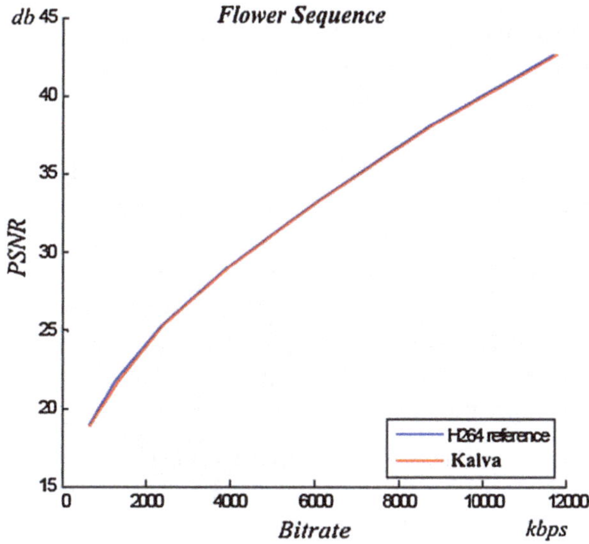

Fig. 3.3 Kalva's transcoder RD performance on *Flower sequence*

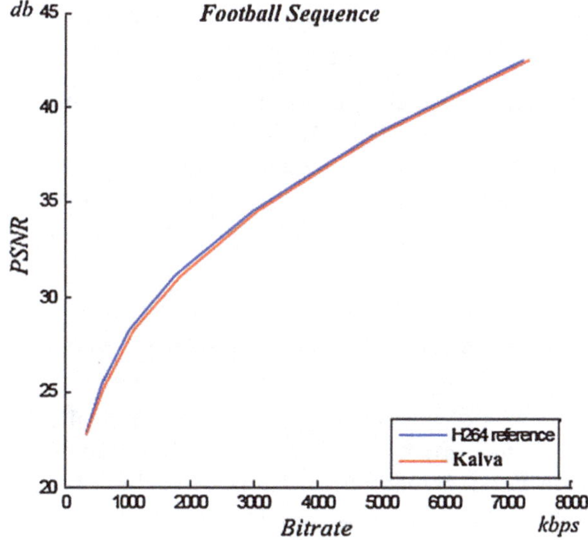

Fig. 3.4 Kalva's transcoder RD performance on *Football sequence*

The experiments, which are conducted with the JM 8.5 of the H.264/AVC standard baseline profiles, show that their approach results in only 30 % drop in the intra frame encoding time with a negligible drop in PSNR. Likewise, the following two sections describe in some details the pros and cons of two different approaches that partially succeed to enhance the computational complexity of the MPEG-2 to H.264/AVC transcoder.

3.2 Kim's Enhanced MPEG-2 to H.264/AVC Transcoding Algorithm

A very interesting MPEG-2 to H.264/AVC enhanced transcoding algorithm was introduced in [8], which will be entitled "Kim's algorithm" in the scope of this book. Liu's solution focused on enhancing the Intra mode decision for the MPEG-2 to H.264/AVC transcoder, which is based on the following DCT characteristics and image processing properties:

- The DCT coefficients of the MPEG-2 decoder in the DCT domain, can be used to compute the spatial variation of the pixel values in the pixel domain.
- Smooth macro-blocks, which have little pixel-value variations, have a higher probability to be encoded in Intra 16×16 mode rather than in Intra 4×4 mode.
- If the 16×16 macro-block has a strong spatial correlation in some direction, then all its 4×4 sub macro-blocks follow the same direction.
- The reconstruction direction decisions for the macro-block's Chroma components are very similar to that of the corresponding macro-block's Luma component as they both represent the same object.

For illustration purpose, we constructed the pseudo code in Fig. 3.5 to demonstrate Kim's algorithm in some details. Additionally, in Chap. 4 for comparison and evaluation purposes, Kim's algorithm is implemented using C++ and compared its achievement to our proposed Intra prediction technique. All things considered, Kim's proposed Intra prediction algorithm can be summarized in the following four steps:

3.2.1 Kim's Intra Prediction Mode Decision Algorithm

By estimating the macro-block smoothness directly from the MPEG-2's DCT domain, Kim's algorithm proposed a fast Intra mode selection algorithm for the H.264/AVC side of the transcoder. First, Kim's algorithm computes a cost function, which utilizes all the DCT coefficients from the MPEG-2 decoder side of the transcoder. And, the energy measure (En) is estimated using the Sum of Absolute Values (SAV) of all the MPEG-2's DCT coefficients of the current 8×8 macro-block and the three neighboring 8×8 macro-blocks: the left, the top and the top left macro-blocks, from the MPEG-2 decoder side of the transcoder as illustrated in Eq. (3.1).

$$E_n = \sum_{i,j=0}^{7} |D_n(i, j)| , n = 0, 1, 2, 3 \tag{3.1}$$

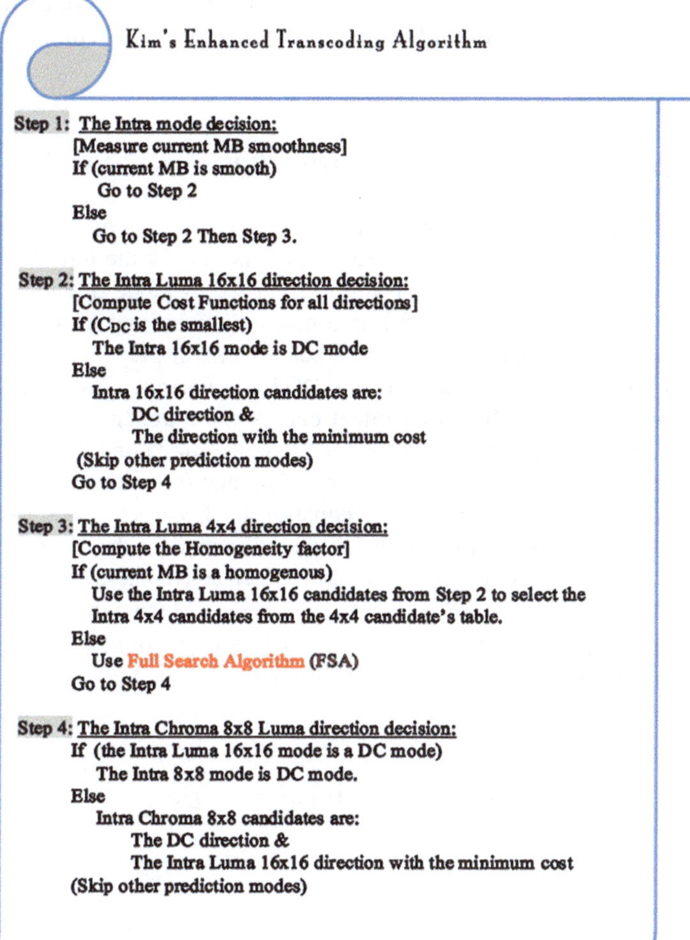

Fig. 3.5 The pseudocode for Kim's transcoding algorithm

Then, the absolute differences of the DCT coefficients are summed in order to compute the cost function (C_{DC}) as shown in (3.2).

$$C_{DC} = 0.5\left\langle \sum_{n=0}^{3} \left|E_n - \bar{E}\right| \right\rangle < \gamma_o \tag{3.2}$$

Afterwards, the current macro-block Luma mode decision is evaluated from the calculated cost function value. On one hand, if the cost function value is more than a threshold (γ_o), then the macro-block is not smooth. Otherwise, the macro-block is smooth and there is no need to evaluate the prediction directions for the Intra 4×4 decision mode, which significantly reduces the Intra Prediction computational complexity.

3.2.2 Kim's Intra Luma 16 × 16 Direction Prediction Algorithm

Furthermore, in case of the current macro-block is a smooth enough macro-block, Kim's algorithm evaluates only the direction prediction for the Intra Luma 16×16 mode. And, it completely skips the Intra Luma 4×4 mode direction prediction. First, the energy of the four 8×8 Luma blocks is measured in the DCT domain using the same four macro-blocks' DCT coefficients in Step (3.2.1). In particular, the Sum of Absolute Values (SAV) of the DCT coefficients is calculated. Also, the SAV has been proved to be an excellent energy measure. In addition, it requires no multiplication, which reduces the process computation complexity. On the other hand, the cost functions proposed in [8] are calculated for vertical, horizontal, plane and DC modes using the following set of equations: 3.3, 3.4, 3.5, and 3.2, respectively. Furthermore, if the cost function of the DC direction mode is the smallest value, then the reconstruction direction of the current Luma 16×16 macro-block is DC mode. Otherwise, the direction candidates are the DC mode and the mode with the smallest cost function mode. Accordingly, Kim's algorithm is performing two RDOs in the worst case for deciding the reconstruction direction for the Luma 16×16 mode.

$$C_V = |E_0 - E_2| + |E_1 - E_3| \tag{3.3}$$

$$C_H = |E_0 - E_1| + |E_2 - E_3| \tag{3.4}$$

$$C_P = 2min\left\{|E_1 - E_2|, |E_0 - E_3|\right\} \tag{3.5}$$

where C_V, C_H and C_P represent the cost functions of the vertical, horizontal and plane, respectively. The 8×8 DCT energy measures E_n, where n = 0, 1, 2, and 3, are defined in Eq. (3.1).

In brief, Kim's algorithm proved that the DCT based cost functions can perfectly detect the best Luma 16×16 prediction direction. Also, Kim's experiments and statistical analysis detected that the computed cost function for the DC mode is usually the smallest value, which indicates that the DC mode is frequently selected as the reconstructing direction for the Intra Luma 16×16 mode. Depending on such experimental results analysis, a fast Intra Luma 16×16 direction prediction method is proposed in [8].

3.2.3 Kim's Intra Luma 4 × 4 Direction Prediction Algorithm

On the other hand, in case of the current macro-block is not a smooth enough macro-block, Kim's algorithm also evaluates the direction prediction for the Intra Luma

Table 3.1 Kim's Intra Luma 4 × 4 direction prediction candidates

Intra 16 × 16 mode	Intra 4 × 4 candidates
0 (V)	0, 2, 5, 7
1 (H)	1, 2, 6, 8
2 (DC)	0, 1, 2, 3, 4
3 (P)	0, 1, 2, 3, 4

16 × 16 mode using the same technique as in Step (3.2.2). Then, it used the best Intra Luma 16 × 16 reconstruction direction to evaluate the direction prediction candidates for the Intra Luma 4 × 4 mode.

In other words, Liu's team proposed a methodology to decide the Intra 4 × 4 direction prediction, which is based on the similarity between the Intra 16 × 16 decision mode and the Intra 4 × 4 prediction mode. First, the homogeneity of the currently encoded macro-block is estimated using their proposed homogeneity factor, which utilizes all the DCT coefficients from the MPEG-2 encoder as in Eq. (3.6):

$$\gamma = \frac{C_{max} - C_{min}}{C_{max}} < \gamma_1 \tag{3.6}$$

$$C_{max} = max\,\{C_x, x = V, H, P\&DC\} \tag{3.7}$$

$$C_{min} = min\,\{C_x, x = V, H, P\&DC\} \tag{3.8}$$

where (γ) is Liu's proposed homogeneity factor and (γ_1) is a constant, which represents a quantization based threshold [8]. C_{DC}, C_V, C_H, and C_P are computed as in Eqs. (3.2, 3.3, 3.4) and (3.5), respectively.

On one hand, if the current 4 × 4 macro-block is homogeneous, then the Intra 4 × 4 direction candidates are dependent on the Intra direction decision of the inclusive 16 × 16 macro-block. And, it can be directly selected from Table 3.1. On the other hand, if the current 4 × 4 macro-block is not homogeneous, then Kim's algorithm falls back to the traditional full search technique.

3.2.4 Kim's Intra Chroma 8 × 8 *Direction Prediction Algorithm*

In [8], the authors investigated the probability that both the Intra Luma 16 × 16 mode and the Intra Chroma 8 × 8 mode are using the same reconstruction direction as both represent the same objects. Furthermore, as illustrated in Fig. 3.6, they discovered that such probability is very high, i.e. over 55 %. In addition, the statistical analysis proved that in case of both the Intra Chroma 8 × 8 and the Intra Luma 16 × 16 are using different prediction directions, the probability of using DC mode for the Chroma 8 × 8 is very high (over 60 %). From such statistics and observations, Liu's team proposed a fast Intra Chroma 8 × 8 direction prediction algorithm that uses no arithmetic operations. It uses the reconstruction direction of the inclusive Luma 16 × 16

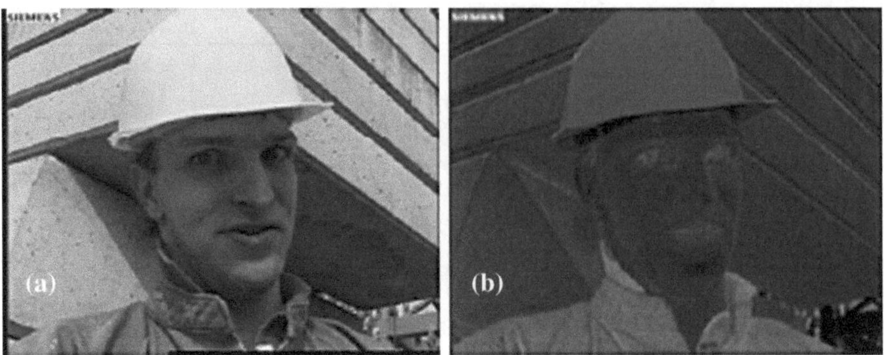

Fig. 3.6 Luma (**a**) and Chroma (**b**) components of *Foreman* frame

macro-block to predict the reconstruction direction for all the four corresponding Chroma 8 × 8 macro-blocks.

All in all, as illustrated by Fig. 3.5, Kim's algorithm utilizes the direction correlation between the 8 × 8 DCT coefficients from the MPEG-2 side of the transcoder to propose a low complexity Intra prediction algorithm for the H.264/AVC side of the transcoder. The algorithm utilizes all the DCT coefficients from each four MPEG-2's 8 × 8 macro-blocks to compute the smoothness of the corresponding H.264/AVC's 16 × 16 macro-block. First, the algorithm decided the Luma operational mode, either the 16 × 16 or the 4 × 4. Then, it nominates the direction prediction candidates for the Luma 16 × 16 mode based on a DCT based cost functions, which also utilize the corresponding MPEG-2's DCT coefficients. Additionally, Kim's Algorithm uses the correlation between the Intra Luma 16 × 16 and the Intra Luma 4 × 4 modes to propose a fast Intra Luma 4 × 4 direction prediction technique. Finally, the authors use the spatial similarity between the Luma and the Chroma components to propose a fast Intra Chroma 8 × 8 direction prediction technique for the H.264/AVC encoder part of the Conventional Cascaded Transcoders (CCT).

Unfortunately, Kim's algorithm fails to avoid using the full search technique for the Luma 4 × 4 mode in case of a non-homogeneous macro-block. And, it had to apply such computational intensive technique in order to maintain the overall coding efficiency. Also, Kim's team used unnecessary DCT coefficient in its calculations, i.e., the DCT's AC coefficients. Furthermore, Kim's algorithm failed to directly predict a reconstruction direction for any of the prediction modes. It only reduces the number of direction prediction candidates for some modes.

3.3 Yoo's Enhanced MPEG-2 to H.264/AVC Transcoding Algorithm

Equally important, an early Intra mode decision method for the H.264/AVC encoder side of the MPEG-2 to H.264/AVC CCT transcoder is proposed in [9]. As well, their algorithm is based on the spatial activity analysis in the video frame level. And, according to the authors' experimental results, they noticed that if the macro-block is not coarse, i.e., it has a little information, they could code such macro-blocks using the larger block size, which is the 16×16 macro-block prediction mode. Otherwise, they have to code the 16×16 macro-block more precisely by dividing it into smaller sub macro-blocks, i.e. the 4×4 macro-block prediction mode. Because such macro-blocks have a high probability to have an important information that must be carefully coded.

In addition, Yoo's team used the DCT coefficients directly from the MPEG-2 decoder in their proposed coding mode judgement technique. However, they used all the DCT's AC coefficients of the current macro-block and the neighboring macro-blocks directly from the MPEG-2 decoder side of the transcoder in order to estimate the energy of the current macro-block. Moreover, they used only the chroma components as shown in Eq. (3.9).

$$Energy_{MB} = \sum_{z=0}^{3} \sum_{i=1,j=1}^{7,7} abs\left(DCT\,[z]\,[i]\,[j]\right) \tag{3.9}$$

where z presents the index of 8×8 sub macro-blocks within the current 16×16 macro-block. And i and j express the location of DCT's AC coefficients for each row and column, respectively.

By using all the Chroma DCT's AC coefficients, the authors computed the energy of the current 16×16 macro-block. They claim that their proposed energy measurement can reflect the spatial activity of the current macro-block. In particular, if the energy is less than a pre-defined threshold (α), then the current macro-block has a little content and it tends to be a smooth macro-block. Therefore, they can select the Intra 16×16 prediction mode as the only prediction candidate and skip all the Intra 4×4 prediction mode calculations. Otherwise, the Intra 4×4 prediction mode is the only prediction mode candidate and they skip all the Intra 16×16 prediction mode calculations.

Furthermore, Yoo's team realizes that the definition of the threshold (α) is the most important step. As the accuracy of the threshold will lead the correctness of the estimated mode. Based on the analysis of their experimental results, they defined the threshold as a Quantization Parameter (QP) based threshold. In other words, they defined the threshold through statistical analysis based on the QP. Figure 3.7 shows the statistical relationship between QP and the threshold (α) for the video sequence "Foreman" as Yoo's team presented in [9].

Even after using the mode skip technique, Yoo's team found out that the complexity of the Intra prediction is still unacceptably high for their targeted applications. For that reason, they proposed a direction prediction method that employs the edge orientation concept to predetermine the direction prediction for the macro-block's Luma and Chroma components.

First, they pre-determined the Intra prediction for each 8×8 Chroma macro-block directly in the DCT domain of MPEG-2 decoder side of the transcoder as shown in Eq. (3.10).

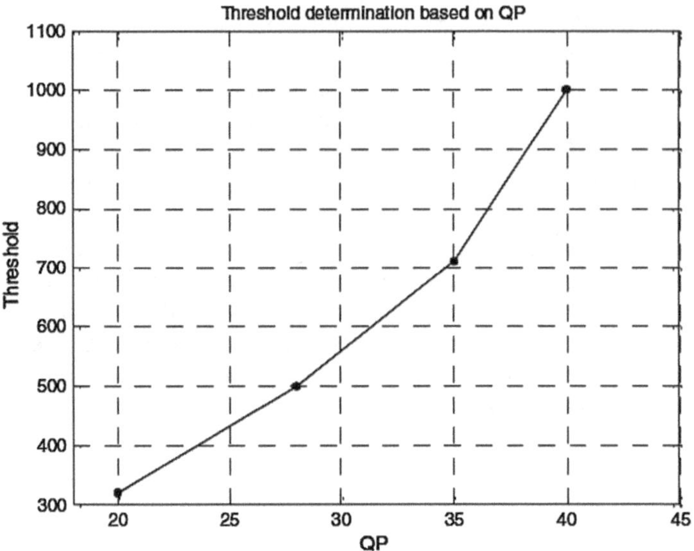

Fig. 3.7 The QP and threshold (α) relationship in *Foreman*

$$\theta = arctan\left(\left(\frac{ENERGY_{left}[comp]}{ENERGY_{top}[comp]}\right)/2\right) \tag{3.10}$$

where $(ENERGY_{left})$ and $(ENERGY_{top})$ are computed as in Eqs. (3.11) and (3.12), respectively.

$$ENERGY_{left} = \sum_{i=8,j=0}^{i<=56,i+=8} DCT[i][j] \tag{3.11}$$

$$ENERGY_{top} = \sum_{i=0,j=1}^{j<=7,j++} DCT[i][j] \tag{3.12}$$

Also, the output value (θ), which falls between ($-\pi/2$) and ($\pi/2$), is used to roughly select the Chroma Intra 8×8 direction prediction candidate for H.264/AVC encoder part of the transcoder. In addition, they decided to add the DC to each predicted direction to propose two candidates Intra direction prediction for each 8×8 Chroma macro-block. Accordingly, from their statistical analysis, they found that the DC prediction is the most frequently used mode.

Using such technique, each Chroma Intra prediction will only run the RDO for two direction prediction candidates instead of the four direction candidates in the H.264/AVC standard Chroma Intra prediction. Moreover, they used the selected

Table 3.2 Intra Luma 4 × 4 direction prediction candidates

Chroma 8 × 8 direction	Luma 4 × 4 direction candidates
1 (V)	0, 2, 5, 7
2 (H)	1, 2, 6, 8
3 (P)	0, 1, 2, 3

Chroma direction candidates as the reconstruction candidates for the corresponding Luma 16 × 16 macro-block without adding any further computations.

On the other hand, in case of the Luma Intra 4 × 4 mode has been selected as the prediction mode for the current macro-block, Yoo's algorithm uses the edge orientation information of Luma 8 × 8 macro-block from the DCT-domain of the MPEG-2 decoder. Besides, it uses similar calculation as in the case of the Chroma components, but with employing the Luma coefficients from the MPEG-2 part of the transcoder.

Additionally, they used a different mode mapping groups, which are summed in Table 3.2. By using such method for the Luma 4 × 4 Intra direction prediction, only four reconstruction direction candidates will be tested instead of the standard nine direction prediction candidates, which currently must be tested in order to decide the best reconstruction direction in the H.264/AVC standard Intra prediction process.

All in all, Yoo's proposed algorithm used the DCT's AC coefficients directly from the MPEG-2 part of the transcoder to pre-select both the operation mode and the reconstruction direction for the Intra prediction process within the H.264/AVC part of the transcoder. After studying and implementing Yoo's algorithm, we have summarized it in the flowchart in Fig. 3.8. Furthermore, we have noticed the following drawbacks:

- **Yoo's Algorithm uses all DCT's AC coefficients**, which are unnecessary increasing the required number of the arithmetic operations to evaluate the energy of the current macro-block as shown in Eq. (3.9).
- **Yoo's macro-block's energy measure is evaluated from the macro-block's Chroma components**, then recomputed from the macro-block Luma component in case of the Luma 4 × 4 mode, which introduces unnecessary redundant computations.
- **Yoo's calculations are neglecting the DCT's DC coefficient**, which contains most of the macro-block energy, and it uses only the AC coefficients, which could be very misguiding and may lead to imprecise prediction.
- **Direct mapping of the current Chroma 8 × 8 macro-block direction prediction candidates** to the corresponding Luma 16 × 16 macro-block is not conceptually correct. As, the edge orientation angle value could dramatically change outside the current Chroma 8 × 8 macro-block while it is still inside the current inclusive Luma 16 × 16 macro-block. In general, using the Chroma components as an indicator for the Luma component is very tricky. However, the Luma components could be precisely used to indicate the corresponding Chroma components.

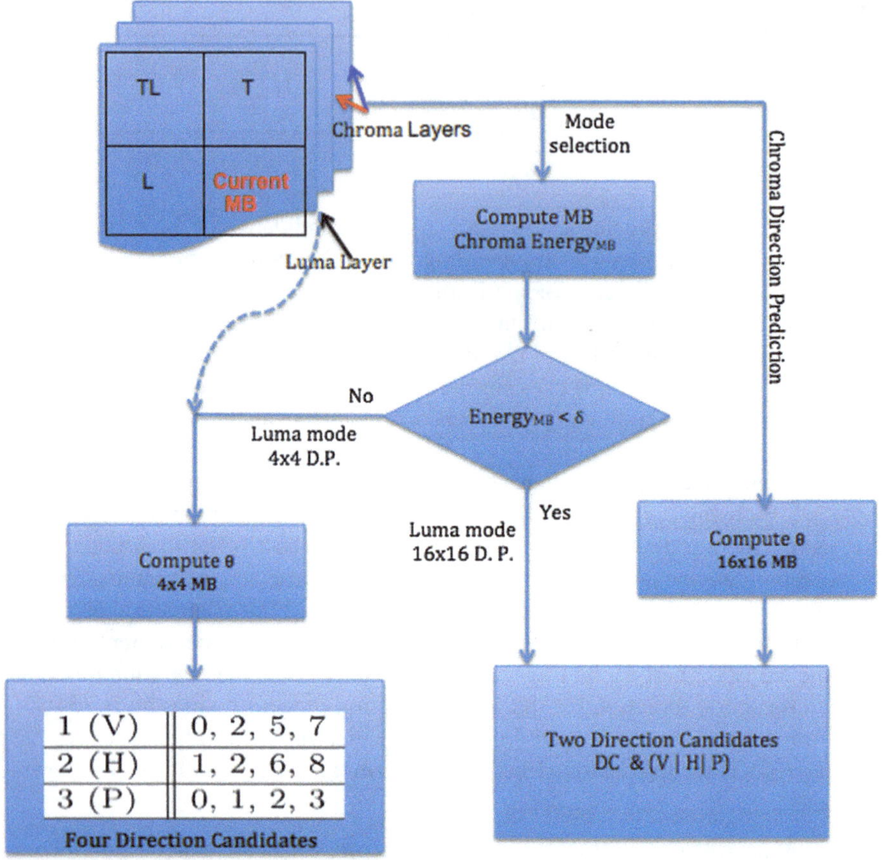

Fig. 3.8 The pseudocode for Yoo's transcoding algorithm

In addition, Yoo's proposed algorithm suffers a high reduction in the PSNR. Accordingly, it introduces an unacceptable degradation in the visual quality. In Chap. 5, we compared both Yoo's achieved Intra prediction run-time and PSNR cost with our proposed Full-search Free (FSF) algorithm Intra prediction run-time and PSNR objective quality, respectively. Also, Yoo's team admitted in [9] the increase in the bit-rate as a pay off for the targeted reduction in the Intra prediction run-time.

In fact, Yoo's solution for addressing the transcoder complexity problem was very motivating to our work presented in this book. They were the first to introduce the concept of using immature information from the MPEG-2 decoder directly into the H.264/AVC encoder in order to accelerate the overall transcoding time. Also, the idea of using the edge orientation for nominating a reconstruction direction to the Intra Prediction process in the H.264/AVC part of the transcoder was very inspiring to our proposed solution for the MPEG-2 to H.264/AVC transcoder complexity problem.

Chapter 4
Real-time MPEG-2 to H.264/AVC Transcoding

Abstract In this chapter, a significantly fast MPEG-2 to H.264/AVC transcoder will be introduced. The proposed efficient transcoder takes advantage of the information stored in the MPEG-2 bit-stream, i.e. the DCT coefficients, to accelerate the Intra prediction process in the H.264/AVC side of the transcoder. While MPEG-2 side of the transcoder does not employ Intra prediction method, the H.264/AVC side eliminates spatial redundancy by the mean of the Intra frame prediction technique. Hence, both the mode selection process and the direction prediction process in the H.264/AVC encoder are responsible for a non-negligible portion of the processing delay in the transcoder. For that reason, we propose a dramatically fast Intra mode selection and direction prediction algorithm, which has been designed and optimized specially for real-time MPEG-2 to H.264/AVC transcoders that are targeting to be used in mobile applications. The proposed MPEG-2 to H.264/AVC transcoding solution proposes an early Intra mode decision and direction prediction techniques, which is based on the spatial activity analysis using our novel smoothness determining factor and our optimized edge orientation detector. Moreover, the proposed fast Intra frame prediction algorithm utilizes the DCT coefficients directly from the MPEG-2 side of the transcoder to eliminate the possible redundant operations and efficiently produce the Intra prediction mode and the macro-block reconstruction direction for the H.264/AVC side of the transcoder. The Experimental results, using the H.264/AVC standard benchmark frames, show 92 % transcoding run-time elimination at the cost of a slight PSNR degradation (less than 3.5 %) compared with the standard conventional cascaded transcoder.

4.1 Introduction

The MPEG-2 compression standard has been widely deployed not only in very popular video applications such as DVDs and DVRs, but also in the video distribution infrastructures like Digital Television (DTV) through both cables and satellites.

T. Elarabi et al., *Real-Time Heterogeneous Video Transcoding for Low-Power Applications*, DOI: 10.1007/978-3-319-06071-2_4,

Additionally, for many years end-to-end MPEG-2 codec systems have existed with millions of interoperable encoders, decoders and multiplexers deployed.

Fortunately, the H.264/AVC standard has been a break through in the area of video codec. The relatively new standard offers a very high compression rate, while the quality degradation cost is preserved to the minimum [7]. For this reason, it has been widely implemented in most of the mobile multimedia devices. In addition, the H.264/AVC is promised to be the universal video codec for all network dependent multimedia applications, e.g. PMP, DVB and Internet video applications because of its practically proved high network adaption capability. Furthermore, the MPEG-2 bit-stream requires almost double the transmission bit-rate and larger storage capacity compared to what H.264/AVC consumes for the same video quality [5]. For that reason, H.264/AVC became the perfect solution for the limited memory of the mobile devices.

Yet, the MPEG-2 to H.264/AVC transcoder is the only reasonable solution for streaming the archived video resources in real-time over the current imperfect networks, e.g. the Internet. Since, most of these valuable video resources have been already encoded in MPEG-2 format for over a decade. Also, the need for MPEG-2 to H.264/AVC transcoding has been recently increased due to the fast spread of multimedia enable mobile devices. Another reason is the large number of the MPEG-2 video libraries that have been already accumulated over the years. In order to access such MPEG-2 videos from devices with different network capabilities, the industry's demand for an efficient fast MPEG-2 to H.264/AVC transcoder has been arisen.

Indeed, the simplest transcoding method is the one that converts the MPEG-2 bit-stream to the H.264/AVC bit-stream by means of cascade transcoding. The idea behind such heterogeneous transcoder is based on a straight forward sequential process that first completely decodes the MPEG-2 bit-stream by using the MPEG-2 standard decoder, then re-encodes the reconstructed video file to H.264/AVC bit-stream by employing the H.264/AVC standard encoder. Moreover, the Conventional Cascaded Transcoder (CCT) is a direct well known, yet naive, MPEG-2 to H.264/AVC transcoding solution. As shown in Fig. 4.1, The CCT has a quite implementation friendly structure, which is achieved only by cascading the MPEG-2 decoder and the H.264/AVC encoder. Although the CCT represents the upper limit on the computational complexity, it theoretically maximizes the video quality [5]. In Chap. 2 of this book, the cascaded MPEG-2 to H.264/AVC transcoder architecture is discussed in some details. In contrast, a computationally efficient transcoder would take advantage of the information stored in the MPEG-2 decoder to accelerate the encoding process in the H.264/AVC encoder side of the transcoder [14]. Indeed, the need for such transcoder presents many research challenges. In fact, many researchers state that the main problem in the MPEG-2 to H.264/AVC transcoding came from the differences between H.264/AVC and MPEG-2 [6]. For example, the MPEG-2 side of the transcoder does not employ the Intra frame compression technique. In contrast, the H.264/AVC encoders apply the Intra prediction compression technique to remove the frame's spatial redundancy. In Chap. 2 of this book, the difference between the MPEG-2 standard formate and the relatively new H.264/AVC formate is introduced in details.

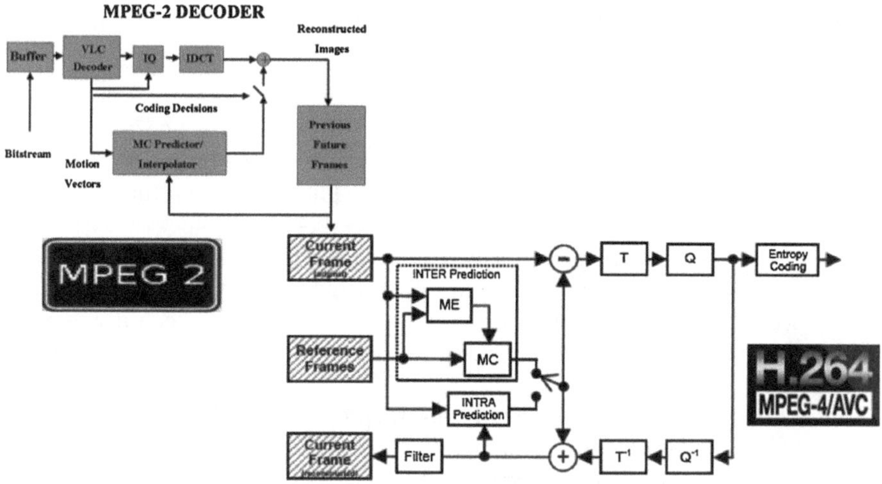

Fig. 4.1 The standard architecture of the MPEG-2 to H.264/AVC CCT

In this chapter, a significantly efficient accelerating algorithm for MPEG-2 to H.264/AVC transcoder is presented. It effectively utilizes the information stored in the MPEG-2 bit-stream, i.e. the DCT coefficients, to reduce the computational complexity of the Intra prediction process in the H.264/AVC encoder side of the transcoder. Specifically, both the mode selection and the direction prediction processes of the H.264/AVC encoder side of the transcoder are responsible for a huge portion of the overall transcoding run-time delay [17]. For that reason, we proposed an early Intra mode selection and direction prediction algorithm, which has been designed and optimized specially for the real-time MPEG-2 to H.264/AVC transcoder. Moreover, the proposed MPEG-2 to H.264/AVC transcoder accelerating solution is based on the spatial activity analysis using both a novel smoothness determining factor and an optimized edge orientation measurer. Ultimately, it proposes an early Intra mode decision and direction prediction techniques for the H.264/AVC side of the transcoder. In particular, the proposed fast Intra frame prediction algorithm utilizes the DCT coefficients directly from the MPEG-2 side of the transcoder to eliminate the redundant operations and efficiently decide the Intra prediction mode and the macro-block reconstruction direction for the H.264/AVC side of the transcoder. The Experimental results, using standard benchmark frames show 92 % transcoding time saving at the cost of a slight PSNR degradation (less than 3.5 %) compared with the standard conventional transcoder.

4.2 High-Throughput MPEG-2 to H.264/AVC Transcoding Algorithm

In Chap. 3, Kim's algorithm performs extensive calculations to decide both the Intra prediction mode and the Intra reconstruction direction for the H.264/AVC encoder side of the transcoder. Furthermore, in the case of the Intra 4x4 mode, Kim's algorithm needs to apply the full-search technique to decide the 4x4 macro-block reconstruction direction in a non-negligible percentage of the macro-blocks of the Intra Frames. On the other hand, Yoo's proposed algorithm suffers a reduction in the PSNR that leads to an unacceptable degradation in the visual quality, which has been discussed in Chap. 3. Moreover, the bit-rate is increased as a pay off for the gained reduction in the Intra prediction run-time.

Motivated by Yoo's and Kim's drawbacks, we proposed an efficiently fast MPEG-2 to H.264/AVC transcoder accelerating algorithm, which significantly reduces the computation complexity of the Intra direction process in the H.264/AVC part of the transcoder. In addition, the proposed transcoding algorithm completely avoids using the full-search technique, which is currently employed by the standard Intra prediction process in order to come up with the best macro-block reconstruction direction for the H.264/AVC decoder side of the CCT transcoder. Successively, the proposed transcoding acceleration algorithm preserves almost the same Peck Signal to Noise Ratio (PSNR) as the current H.264/AVC reference software is offering.

The flowchart in Fig. 4.2 briefly summarizes our proposed transcoder accelerating algorithm. Accordingly, the proposed transcoding solution processes the video frames in units of four 8×8 macro-blocks from the MPEG-2 side of the transcoder, which are corresponding to one 16×16 macro-block in the H.264/AVC side of the transcoder. All things considered, the proposed algorithm includes two phases, the mode decision phase and the direction prediction phase.

4.2.1 The Intra Mode Decision Phase

The first phase in the proposed accelerated transcoding algorithm is deciding the H.264/AVC Intra prediction mode. In other words, in this phase the algorithm decides if either, it can select the prediction direction for the whole 16×16 macro-block as one unit or, it needs to be more precise by dividing the 16×16 macro-block into sixteen non-overlapping 4×4 sub macro-blocks. The proposed decision technique is built upon an optimized version of the smoothness factor that was first introduced by Kim's transcoding algorithm in [8]. Similarly, our proposed smoothness factor is computed from the energy of the current macro-block discrete cosine transform (DCT) coefficients directly from the MPEG-2 side of the transcoder. However, only the DCT' CD coefficients of four non overlapping 8×8 macro-blocks from the MPEG-2 decoder is used in the smoothness factor calculations as modelled in Eqs. (4.1), (4.2) and (4.3). Furthermore, if comparing the optimized smoothness

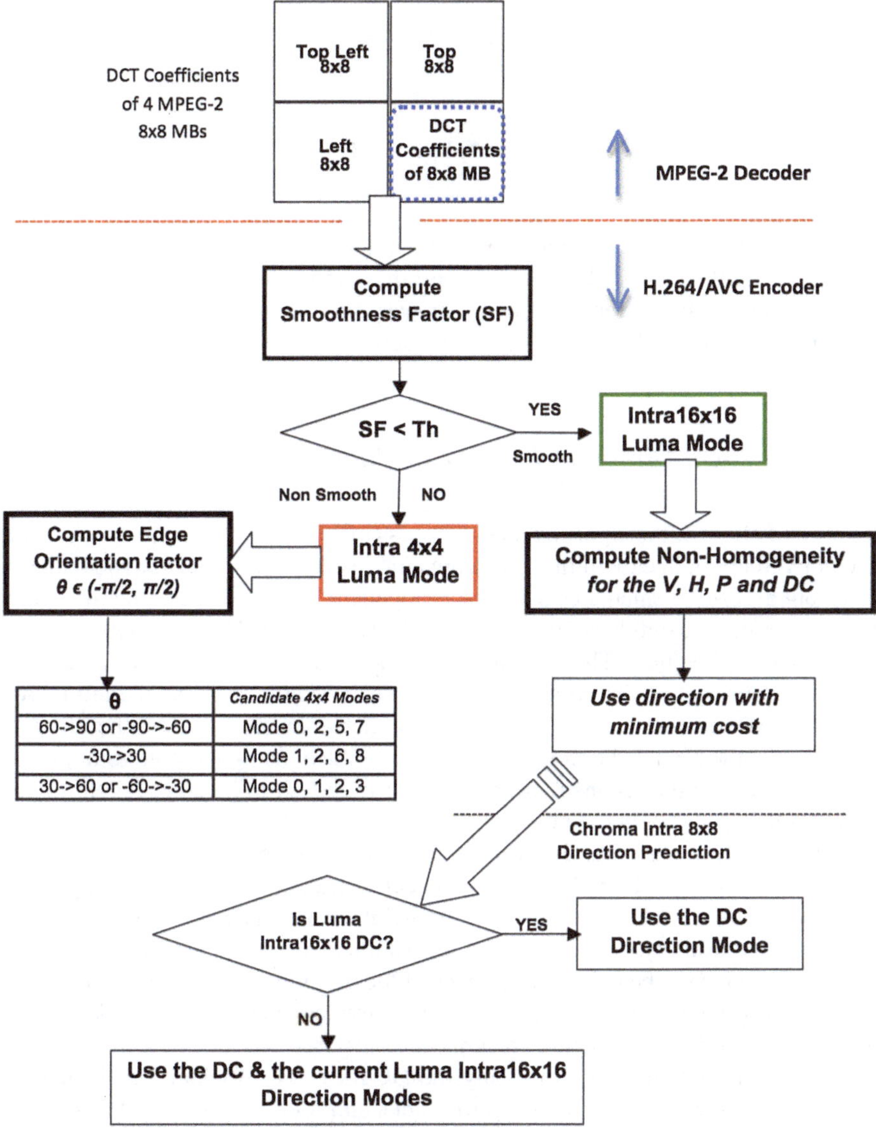

Fig. 4.2 The proposed MPEG-2 to H.264/AVC transcoding algorithm

factor to the QP threshold indicates a smooth enough macro-block, then the mode decision is set to the 16 × 16 mode. Otherwise, the mode decision is set to the 4 × 4 mode.

$$E_n = DCT_n(0,0),\ n = 0, 1, 2 \text{ and } 3 \tag{4.1}$$

$$\bar{E} = \left\{ \sum_{n=0}^{3} E_n \right\} \div 4 \tag{4.2}$$

$$SF = \sum_{n=0}^{3} \left| E_n - \bar{E} \right| \tag{4.3}$$

where E_n refers to the estimated energy of the nth 8×8 sub macro-block. $DCT_n(0, 0)$ is the DCT's DC value of the nth macro-block. $\bar{E}$ is the average estimated energy of the current four 8×8 macro-blocks. And, SF is the estimated smoothness factor for the current 16×16 macro-block.

4.2.2 The Intra Direction Prediction Phase

The second phase of the algorithm is the process of selecting the macro-block reconstruction direction. In case of the Intra Luma mode decision is the 16×16 prediction mode, our algorithm calculates the non-homogeneity factor for all of the four Luma 16×16 direction candidates: the Vertical (V), Horizontal (H), Plane (P) and (DC) reconstruction direction. Then, it employs these factors to decide the Intra Luma 16×16 reconstruction direction. The decision is based on choosing the Intra direction that performs the least non-homogeneity factor. In other words, the direction that achieves the highest homogeneity is chosen. Also, using the non-homogeneity measure to decide the reconstruction direction is proved to be an excellent logical methodology specially in case of the smooth Intra frames [8].

Furthermore, in case that the current macro-block pixels are homogeneous in a specific direction, which implies that the objective frame within the currently examined macro-block is extended through the neighboring macro-blocks at the same direction. Accordingly, the values of the examined macro-block pixels could be fairly predicted from the neighboring macro-blocks at the direction that achieves the highest homogeneity. The non-homogeneity factors calculation process is inherited from Kim's algorithm, but with two enhancements.

The first enhancement is using only the DCT's DC coefficient to calculate the energy of the current four 8×8 macro-blocks and totally excluding the AC coefficients from the calculations as shown in Eq. (4.1). Such modification significantly reduces the required number of arithmetic operations. Consequently, it reduces the overall computational complexity of the MPEG-2 to H.264/AVC transcoder.

The second improvement is the computations of the non-homogeneity factor for the plane direction as shown in Eq. (4.7). The following set of modelling equations show how the non-homogeneity factors are calculated in the proposed enhancing transcoding algorithm for each of the Intra Luma 16×16 direction candidates:

$$H_{DC} = SF \div 2 \tag{4.4}$$

Table 4.1 The Intra Luma 4 × 4 direction prediction candidates

The θ value	Luma 4 × 4 direction candidates
60:90 OR −90:−60	0, 2, 5, 7
−30:30	1, 2, 6, 8
30:60 OR −60:−30	0, 1, 2, 3

$$H_V = |E_0 - E_2| + |E_1 - E_3| \quad (4.5)$$

$$H_H = |E_0 - E_1| + |E_2 - E_3| \quad (4.6)$$

$$H_P = |E_0 - E_3| + |E_1 - E_2| \quad (4.7)$$

After calculating the non-homogeneity factors for each of the Intra Luma 16 × 16 reconstruction direction candidates, the direction that achieves the minimum non-homogeneity factor is chosen as the reconstruction direction for the current Luma 16 × 16 macro-block.

On the other hand, in case of the Intra Luma mode decision is 4 × 4 prediction mode, the direction prediction decision is built upon an edge orientation factor. The concept of using the edge orientation in the Intra prediction process was first introduced by Yoo's algorithm in [9]. However, the computation methodology that is used by our transcoding algorithm for estimating the edge orientation factor is completely different than the one introduced in Yoo's algorithm. In the proposed algorithm, the edge orientation angle is estimated from the DCT coefficients of the four 8 × 8 MPEG-2 macro-blocks as shown in Eqs. (4.8), (4.9) and (4.10). Based on this angle, the 4 × 4 direction candidates are chosen from Table 4.1.

In short, the direction candidate that achieves the minimum loss is chosen as the reconstruction direction for the current examined Luma 4 × 4 macro-block. The process of calculating the estimated edge orientation angle is modelled in Eq. (4.8):

$$\theta = arctan\left\langle \left\{ \sum_{n=0}^{3} Eng_{top}(n)/Eng_{left}(n) \right\} /4 \right\rangle \quad (4.8)$$

$$Eng_{top}(n) = \sum_{i=0,j=1}^{j=7,j=j+1} |DCT_n(i,j)| \quad (4.9)$$

$$Eng_{left}(n) = \sum_{i=1,j=0}^{i=7,i=i+1} |DCT_n(i,j)| \quad (4.10)$$

Afterwards, the Intra direction prediction for the current 16 × 16 macro-block Chroma components in the baseline profile is performed without any additional calculations. The proposed algorithm only checks if the current Luma 16 × 16 direction prediction is the DC reconstruction direction. And, if the current Luma 16 × 16

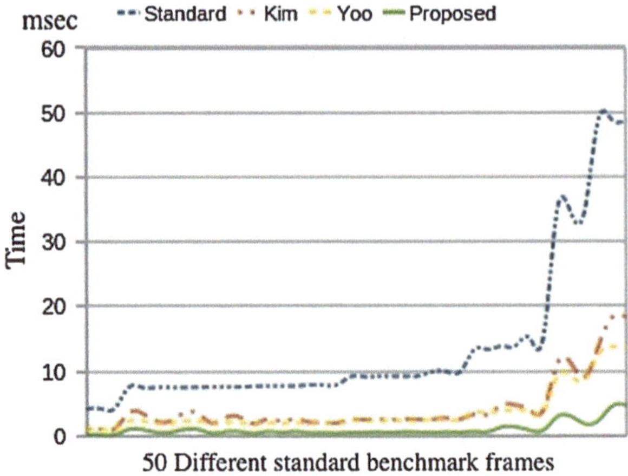

Fig. 4.3 Transcoding run-time for 50 benchmark frames

reconstruction direction is the DC direction, then the Chroma reconstruction direction is the DC direction. Otherwise, the Chroma reconstruction direction candidates will be the DC direction and the current Luma 16×16 reconstruction direction.

4.3 Experimental Results of the Proposed Transcoding Solution

For the sake of testing and evaluation, not only the proposed MPEG-2 to H.264/AVC accelerated transcoding algorithm, but also Kim's and Yoo's algorithms in [8, 9], respectively were implemented using C++. In addition, the Conventional Cascaded Transcoder (CCT) has been implemented using the MPEG-2 standard decoder and the encoder version of H.264/AVC reference software JM 17.2. Furthermore, a comparison between all the implemented algorithms and the standard full-search algorithm that is currently employed by the CCT transcoder has been developed in terms of Peak Signal to Noise Ratio (PSNR) and transcoding run-time.

The system platform, which we have used for our experiments, is Intel Core 2 Due 2.4 GHz CPU, 4 GB 1067 MHz DDR3 RAM and Mac OS X Lion 10.7.2. Moreover, the standard benchmark that has been used to test all the implemented Intra prediction algorithms consists of fifty different classical and modern standard video frames as shown in Appendix [1].

Figure 4.4 shows the run-times needed by each of the implemented transcoding algorithms to completely transcode each frame of the previously described standard benchmark. Even more, Fig. 4.3 is averaging up the results, which have been presented in Fig. 4.4. On the other hand, Fig. 4.6 illustrates the PSNR achieved by each frame of the previously described standard benchmark after it is completely transcoded to H.264/AVC format by each of the implemented transcoding algorithms. In the same way, Fig. 4.5 is averaging up the results, which have been presented in

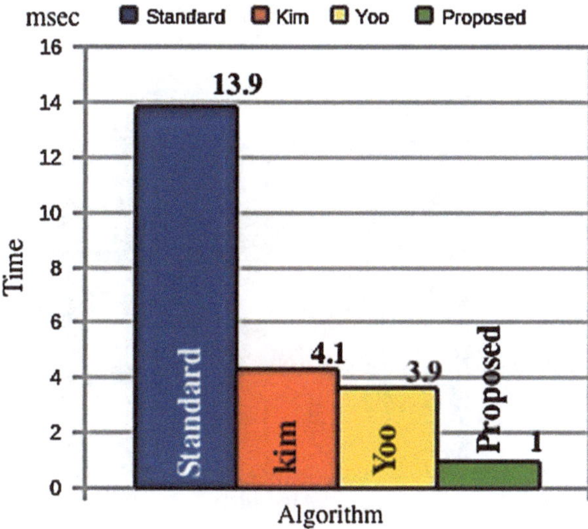

Fig. 4.4 Average run-time for each of the implemented algorithms

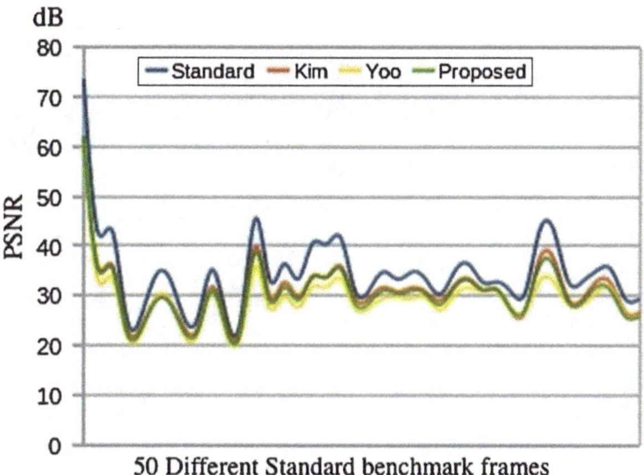

Fig. 4.5 PSNR for 50 benchmark frames

Fig. 4.4. Although the proposed transcoding algorithm introduces a reduction in the PSNR, the visual quality of all the tested frames is comparable to the quality results from using the conventional transcoder.

All in all, the experimental results show that the proposed accelerated transcoding algorithm enhances the transcoding run-time nearly with a factor of 75.6 and 74.4 % when comparing to both Kim's and Yoo's algorithms, respectively. In addition, it achieves 92.8 % transcoding run-time reduction, when comparing to the transcoding run-time that is required by the full-search algorithm in the CCT transcoder.

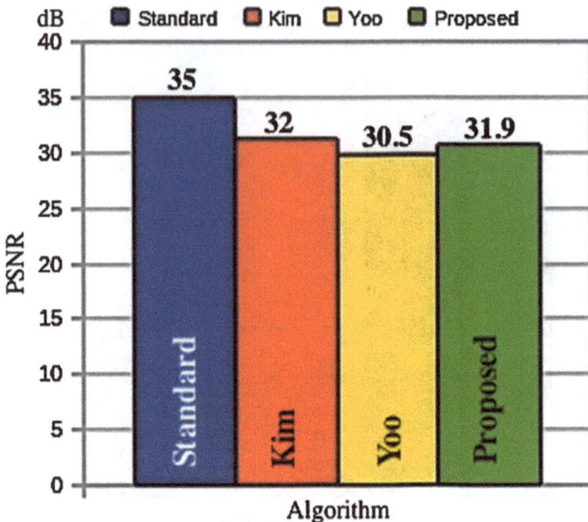

Fig. 4.6 Average PSNR for each of the implemented algorithms

Fig. 4.7 The *building* frame by the CCT transcoder

However, the cost is a slight drop in the PSNR (almost 3.1 dB), less than the standard full-search algorithm in the CCT transcoder. On the other hand, the achieved PSNR is very similar to the PSNR achieved by Kim's algorithm and 4.4 % better than Yoo's algorithm. To demonstrate, Figs. 4.7 and 4.8 are an example for the best case visual

Fig. 4.8 The *building* frame by the proposed transcoder

Fig. 4.9 The *trees* frame by the CCT transcoder

Fig. 4.10 The *trees* frame by the proposed transcoder

quality that we have achieved out of all the tested frames. In contrast, Figs. 4.9 and 4.10 are a sample for the worst case visual quality that we have achieved during our experiments.

4.4 Conclusion

In this chapter, we have presented a fast, yet precise, Intra mode prediction and reconstruction direction algorithm that employs the DCT's DC coefficients directly from the MPEG-2 side of the transcoder for such purpose. To sum up, the proposed accelerated MPEG-2 to H.264/AVC transcoding algorithm achieves an average of 92.8 % reduction in the required transcoding run-time at a price of an acceptable PSNR degradation. Such empirical results nominate the proposed enhanced transcoding algorithm for real-time video transcoding applications.

Indeed, the proposed Intra prediction technique successfully decides the Intra macro-block mode then predicts its reconstruction direction in a completely Full-Search Free technique. The proposed early Intra mode decision and direction prediction method significantly decreases the complexity of the H.264/AVC encoder part of the CCT transcoder. In Chap. 5, the proposed Intra perdition technique will be modified to fit into the standard H.264/AVC decoder.

Chapter 5
Full-Search Free Intra Predication for H.264 Decoder

Abstract The standard H.264/AVC Intra frame encoding process has several data dependent and computational intensive coding methodologies that limit the overall codec performance. It causes not only a high degree of computational complexity, but also an unacceptable delay especially for the real-time video applications. In Chap. 4, we have introduced an enhanced algorithm and its software solution for the Intra prediction in the H.264/AVC encoder side of the MPEG-2 to H.264/AVC transcoder. However, at some point, the information throughput capability of the software solution cannot keep up with the throughput requirement of real-time applications. Even with powerful high speed CPU's, a backlog often occurs during the decoding process. In addition, when split-second situational awareness and decision making require the processing of huge volumes of information in real-time, the real benefit of the hardware solution shines. Moreover, the hardware-based solution introduces a far less parasitic drag than the software systems of the past. Clearly, the need for speed, which is inherited from the real-time demanding applications, necessitates the development of hardware solution for video codec standards. In this chapter of the book, a Low-power hardware architecture for the proposed high throughput Full-Search Free (FSF) Intra mode selection and direction prediction algorithm is proposed. In fact, the proposed Intra prediction hardware solution could be easily integrated in the H.264/AVC encoder side of the MPEG-2 to H.264/AVC transcoder, which has been discussed in Chap. 4. However, we realized that our design can perfectly work as an accelerating hardware solution for the stand alone standard H.264/AVC decoder devices. Accordingly, we focused on modifying the proposed FSF algorithm to fit into the standard H.264/AVC decoders. At first, we tested the modified version of the FSF algorithm after integrating it with the JM 18.2 reference software of the standard decoder. Then, we designed a power efficient hardware architecture for the proposed high throughput FSF algorithm. As a result, the introduced low-power/high-throughput hardware implementation significantly reduces the computational complexity and the processing run-time that are required for the H.264/AVC Intra frame prediction process. As illustrated in Fig. 5.1, the FPGA prototyped version for the proposed architecture has achieved an operating frequency of 291 MHz at a

T. Elarabi et al., *Real-Time Heterogeneous Video Transcoding for Low-Power Applications*, DOI: 10.1007/978-3-319-06071-2_5,

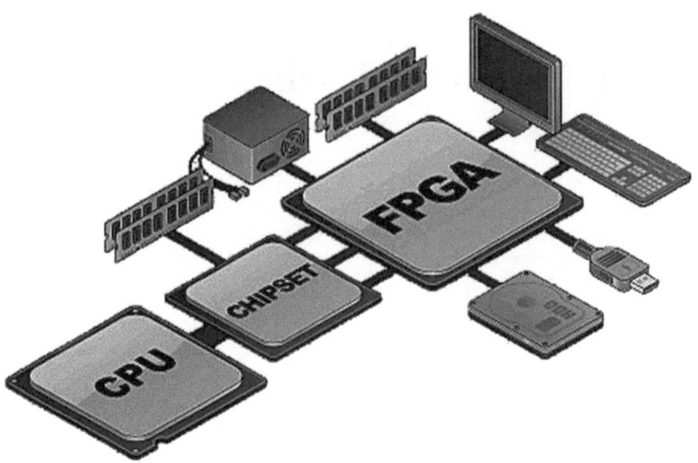

Fig. 5.1 The layout of the FPGA testing system

power consumption of 112 mW when implemented on Virtex-5 FPGA development board. The FPGA simulation and testing that employed a standard video benchmark proved the efficiency and the functionality of the proposed Intra prediction architecture, specially for real-time H.264/AVC video codecs on mobile devices. Motivated by the FPGA experimental results, the ASIC implementation for the FSF architecture is carried out and synthesizing results are obtained. Our heavily tested 45 nm ASIC design was able to achieve an operating frequency of 140 MHz and limiting the overall power consumption to only 9.01 mW, which is strongly nominating our proposed Intra prediction hardware architecture for interactive real-time mobile video applications.

5.1 Introduction

The H.264/AVC video codec standard is also known as MPEG-4 part 10. It is the video coding standard recommendation that is jointly developed and approved by both the Video Coding Experts Group (VCEG) and the Moving Picture Experts Group (MPEG). The joint efforts of both video research groups have produced a breakthrough in the area of video coding. In brief, they were able to achieve double the compression rate while preserving the same visual video quality as the previous standards, e.g. MPEG-2 and H.263.

Afterwards, the H.264/AVC standard has been widely implemented in most mobile multimedia terminals due to its high encoding efficiency at the cost of minimum visual quality degradation. Also, it has shown a high network adaptive capability for error prone networks like mobile channels, which have high tendency for bit errors and packet losses. Accordingly, the H.264/AVC standard is promised to be

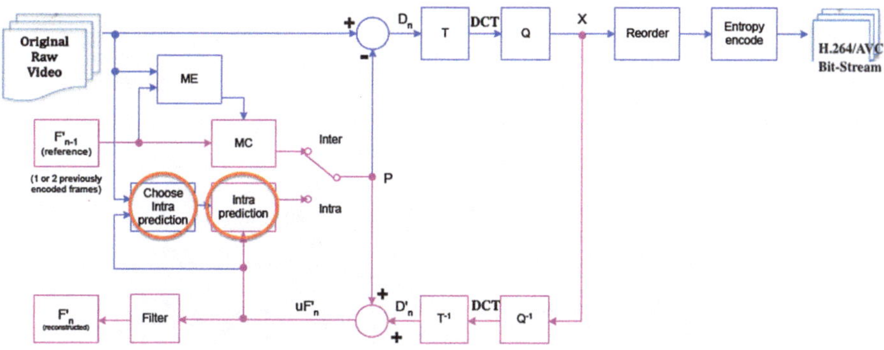

Fig. 5.2 The H.264/AVC standard encoder architecture

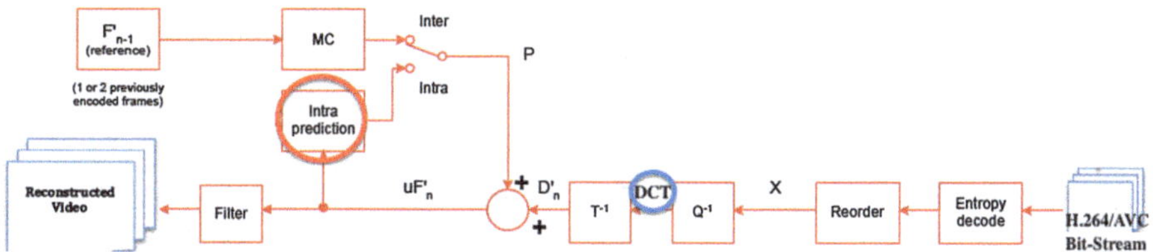

Fig. 5.3 The H.264/AVC standard decoder architecture

the universal video codec for all network dependent video applications, e.g. Digital Video Broadcasting (DVB) and Internet video conferences, etc.

In fact, the H.264/AVC standard employs many new coding tools to improve the coding performance compared to previous standards, which include, but not limited to, the variable block size motion estimation, the integer Discrete Cosine Transform (iDCT) and the Context Adaptive Binary Arithmetic Coding (CABAC) [2]. All these enhancing tools and algorithms enable the H.264/AVC standard to provide high video compression performance for non-interactive applications like video streaming, which requires high coding efficiency at low bit-rate. However, these privileges are achieved at the cost of additional computational complexity, which is caused by the advanced video coding algorithms at the different H.264/AVC operational stages. Figures 5.2 and 5.3 shows the standard building architecture blocks of the H.264/AVC encoder and decoder, respectively [7].

Furthermore, both the empirical and the implementation results have shown that the computational complexity of the H.264/AVC is ten times higher than the MPEG-4 Advanced Visual Profile (MPEG-4/AVP). Therefore, the H.264/AVC codecs have not been able to fulfill the throughput constraints for the real-time interactive applications, e.g. the different video telephony applications. Indeed, the real-time interactive video processing systems require both a high throughput encoder and a low latency decoder.

As a rule, the performance of the standard H.264/AVC Intra frame codec is depending on several computationally intensive coding tools such as the multiple modes Intra

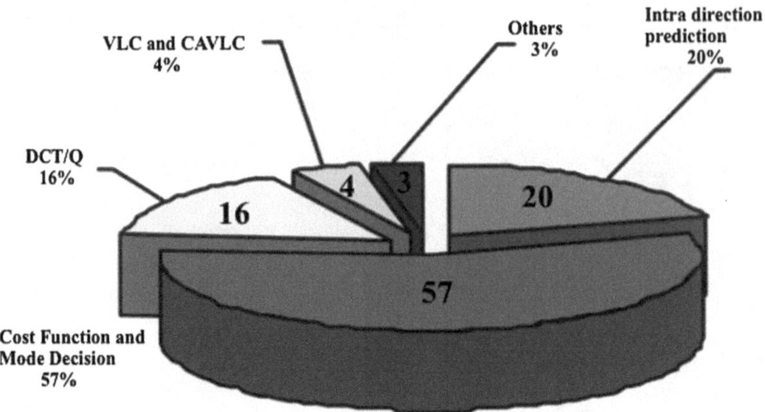

Fig. 5.4 The run-time delay of the H.264/AVC standard codec blocks

prediction, the Lagrangian-based mode selection and the different entropy coding algorithms [3]. In fact, the Intra prediction stage is one of the most critical processes in the H.264/AVC codec as the frame's spatial correlation is used to dramatically improve the overall compression ratio. Also, it plays an important part in rising up the codec's overall performance by significantly reducing the bit-rate. However, it introduces the dominant part of the coding overall computational complexity. Figure 5.4 presents the run-time percentages of the main components of the H.264/AVC stander decoder [17].

Clearly, the cost function generation and the mode decision modules are introducing the worst delay that is causing over half of the decoder's overall run-time cost. Moreover, the Intra Prediction introduces the second worst delay, which is estimated as one fifth of the decoder's run-time cost. All in all, the two previously discussed modules are causing almost 77 % of the computational complexity of the H.264/AVC standard decoder. Indeed, the run-time analysis declares the Intra mode selection and direction prediction as the bottleneck of the H.264/AVC standard codec.

Ordinarily, in the H.264/AVC Intra prediction process, every macro-block is encoded by using neighboring MBs' samples, which have been already encoded within the same frame. The Intra frame coding methodology is comparable to that of the JPEG2000 image compression [4]. Also, the Intra frame encoding, which is currently employed in H.264/AVC video coding standard, is suitable for both image and video compression. However, its high computational complexity has been a barrier for employing the H.264/AVC codecs for interactive real-time video applications.

Successively, several efforts have been introduced to reduce the computational complexity of the H.264/AVC standard and to adapt it for interactive real-time applications. For its dominant contribution in increasing the overall codec computational complexity, a non negligible number of research papers, thesis and dissertations have been focused on enhancing the Intra prediction part of the H.264/AVC. A quite

Fig. 5.5 The FPGA and the ASIC hardware implementations

interesting work has been introduced in [8]. The authors of this paper have proposed, as discussed in Sect. 3.2, a mode-skipping rule for the Intra prediction mode decision in the H.264/AVC encoder. Their idea is based on their experimental results, which show that the DCT energy has a strong correlation within the Intra mode selection and direction prediction in H.264/AVC codec. Additionally, the research team in [9] proposed an Intra decision method based on spatial activity analysis for the DCT coefficients as discussed in Sect. 3.3 in some details.

Furthermore, an interesting H.264/AVC Intra frame algorithm has been proposed in [18]. In particular, the authors proposed to eliminate the "Plane" mode prediction in order to speed up the encoding process. Also, they proposed an enhanced cost function to compensate the video quality. And, they presented their work within three steps heuristic Intra prediction algorithm. In brief, their proposed algorithm skips to predict the modes that have lower probability to be the best mode.

In this chapter, a hardware efficient architecture for the fast yet precise Full-Search Free (FSF) Intra mode selection and direction prediction algorithm in the H.264/AVC decoder is proposed. The heavily tested software implementation for the FSF algorithm selects the macro-block mode, then predicts the reconstruction direction without employing the standard full-search technique that is currently employed by the Intra prediction module in the standard H.264/AVC decoder. Additionally, the proposed FSF low-power hardware architecture is discussed and prototyped using Virtex-5 FPGA development board. Then, motivating by the encouraging results of the FPGA prototyping, the ASIC implementation results for the FSF Intra prediction accelerating architecture is introduced (Fig. 5.5).

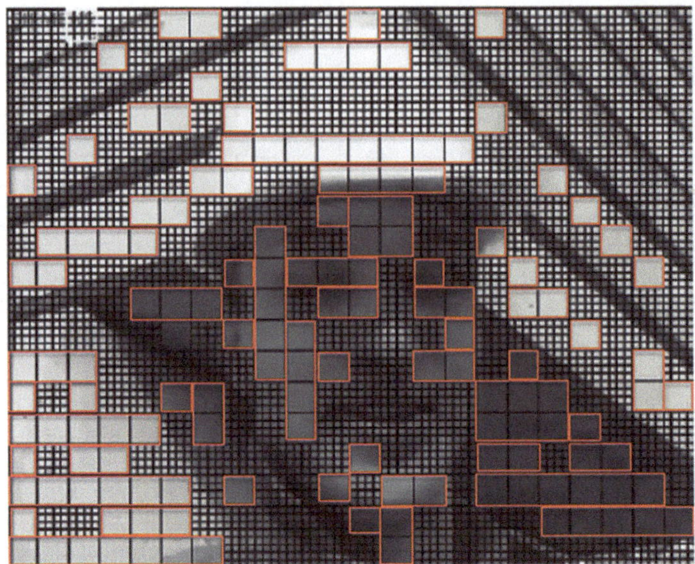

Fig. 5.6 The standard H.264/AVC mode selection for *Foreman*

5.2 The Full-Search Free (FSF) Intra Prediction Algorithm

The Full-Search Free Intra prediction algorithm (FSF) is based on the characteristics of the integer Discrete Cosine Transform (DCT), image processing properties and some other experimental observations. The first observation is the high probability that the smooth macro-blocks are encoded using the Intra 16×16 mode.

As illustrated by Fig. 5.6, the macro-blocks that have low information are rarely encoded using the 4×4 Intra mode. In other words, the smooth macro-blocks are usually encoded using the 16×16 mode. To clarify such observation, we have extracted and highlighted the smooth area in several standard H.264/AVC benchmark video frames. Then, we decoded all the frames using the H.264/AVC JM 18.2 standard encoder while we were monitoring the employed Intra mode that is selected by the standard Intra prediction module.

Furthermore, the smoothness of the examined macro-blocks is estimated according to the pixel value variation within the 256 pixels of each 16×16 macro-block. After all, we contracted the results of the experiment in Table 5.1.

Clearly, the results of the mentioned experiment indicate that the 16×16 Intra prediction mode is the dominant selected mode by the standard H.264/AVC Intra frame process for the smooth macro-blocks, which hold little to no information. Additionally, it has been experimentally proved that in case of the 16×16 Intra prediction mode that has a strong spatial correlation in some direction, all its 4×4 sub macro-blocks follow the same reconstruction direction [8].

Table 5.1 The probability of 16 × 16 mode at different pixel value variation

Frame	Variance					
	0–20 (%)	21–40 (%)	41–60 (%)	61–80 (%)	81–100 (%)	Over 100
Foreman	97	81	60	32	21	6
Cameraman	96	83	56	31	19	9
Peppers	99	87	65	35	22	8
Lena	97	83	61	41	18	7
Mandrill	89	80	49	29	21	5
Jet plane	91	79	50	28	22	6

On the other hand, the analysis of our experiments on classical and modern standard video processing benchmark has showed that the predicted reconstruction directions for both Chroma components are similar to the predicted direction for the corresponding Luma component of the same macro-block. Because, they both represent the same image object or area. Moreover, in the Intra prediction algorithm in [8], the authors used the spatial similarity between the Luma 16 × 16 macro-block component and their corresponding Chroma components to propose a fast Intra mode decision for the Chroma components. In addition, the authors used the correlation between the Intra 16 × 16 mode and the Intra 4 × 4 mode to produce a fast Intra direction prediction decision.

Unfortunately, their algorithm performs extensive calculations for deciding both the Intra macro-block mode and the Intra reconstruction direction. At the same time, in case of deciding the 4 × 4 mode as the current macro-block Intra prediction mode, their proposed algorithm employs the H.264/AVC standard full-search method to decide the 4 × 4 macro-block reconstruction direction in a non-negligible percentage of the Intra frame's macro-blocks. In Sect. 3.2, we discuss in some details their algorithm, which entitled as "kim's algorithm" in the scope of this book. Accordingly, the FSF Intra prediction algorithm is proposed to eliminate such drawbacks in [8]. It completely avoids employing the full-search technique for predicting the Intra mode and the reconstruction direction while preserving almost the same PSNR achieved by the current H.264/AVC JM 18.2 standard software. Consequently, it significantly reduces the computation complexity of the H.264/AVC Intra prediction process.

The flowchart in Fig. 5.7 summarizes the proposed FSF algorithm. In the first place, the FSF algorithm processes the Intra video frame in macro-block units of size 16 × 16 samples. As follows, the algorithm is composed of two phases:

The first phase only concerns with deciding the Intra prediction mode. In other words, it decides the operational macro-block size, either 16 × 16 samples or 4 × 4 samples. Depending on the macro-block smoothness, the FSF algorithm either selects the reconstruction direction for the whole 16 × 16 macro-block as one unit or it divides the 16 × 16 macro-block into its 16 non-overlapped 4 × 4 sub macro-blocks. Then, it decides the best reconstruction direction for each of sub macro-block

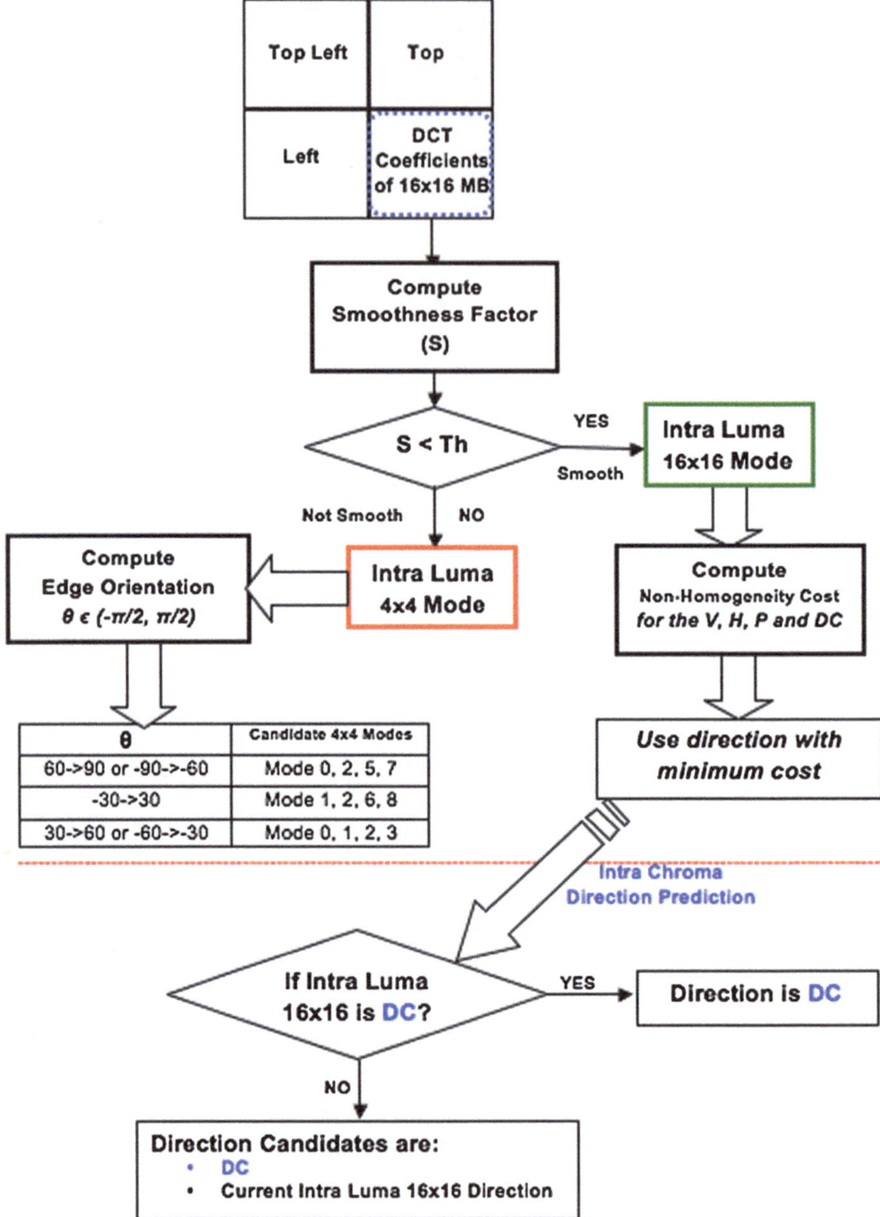

Fig. 5.7 The flowchart of the proposed FSF Intra prediction algorithm

independently. In fact, the FSF mode selection process builds its decision by employing our optimized version of the smoothness factor for the 16×16 macro-block that was first introduced in Kim's algorithm [8].

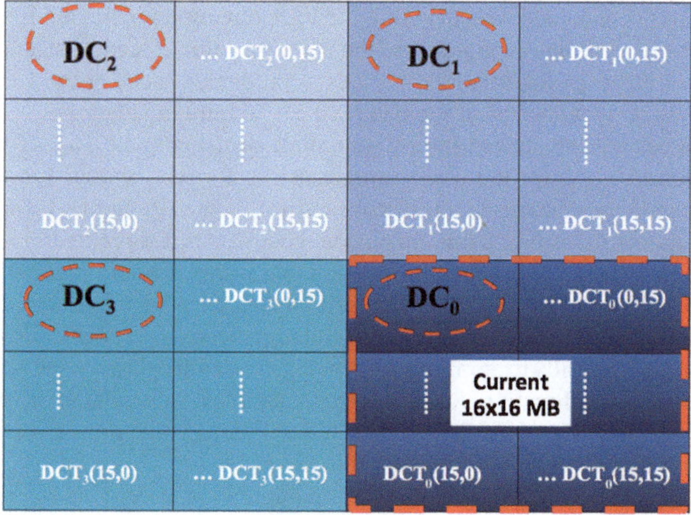

Fig. 5.8 The DCT coefficients used for (S) calculation

In brief, if the optimized smoothness factor indicates a smooth enough macro-block, then the mode decision is the 16×16 mode. Otherwise, the mode decision is the 4×4 mode. In contrast with Kim's algorithm, the FSF mode decision process is using only the DC coefficients from the current macro-block DCT samples to calculate the smoothness factor (S) as illustrated in Eqs. (5.1 and 5.2).

Successively, the process is initiated and fully completed in the DCT domain of the H.264/AVC decoder. As shown in Fig. 5.3, the DCT coefficients are used directly after the H.264/AVC inverse quantization process (Q^{-1}) is completed and before the inverse transformation process (T^{-1}) is initiated.

$$E_{avg} = \left\{ \sum_{m=0}^{3} DC_m \right\} \div 4 \tag{5.1}$$

$$S = \sum_{m=0}^{3} \left| DC_m - E_{avg} \right| \tag{5.2}$$

where DC_m is the DCT's DC sample of the mth macro-block. E_{avg} is the average estimated energy for the four 16×16 macro-blocks: the current macro-block "0", the top macro-block "1", the top/left macro-block "2" and the macro-block directly to the left "3" as shown in Fig. 5.8. Finally, S is our optimized smoothness factor for the current 16×16 macro-block [15].

Afterwards, the second phase of the FSF algorithm is triggered by the output decision of the FSF mode selection phase for the currently processed macro-block.

Also, it is only concerned with deciding the Intra direction prediction. In this phase of the FSF Intra prediction algorithm, the reconstruction direction or the direction candidates are chosen for either the 4×4 Intra prediction mode or the 16×16 Intra prediction mode that have been decided by the FSF Intra mode phase.

First, in case of the current Intra Luma mode decision is the 16×16 mode, the FSF algorithm calculates the non-homogeneity factor for the four Luma 16×16 direction candidates: Vertical (V), Horizontal (H), Plane (P) and DC. Then, it uses the resulting factors' values to nominate one reconstruction direction for the current 16×16 macro-block. It chooses the direction that shows the least non-homogeneity value. In other words, the algorithm chooses the direction that introduces the highest homogeneity. Clearly, if the macro-block is homogeneous in a specific direction, this implies that the currently examined macro-block is extended through the neighboring macro-blocks at that direction. Accordingly, the samples' values of the current macro-block can be precisely predicted from such neighboring macro-blocks' samples at the direction that achieves the highest homogeneity.

Similarly, the non-homogeneity factors computation process is inherited from Kim's algorithm in [8], but with two significant enhancements. The first enhancement is that the FSF algorithm uses only the DC coefficient out of the four adjacent macro-blocks' DCT samples to calculate the current macro-block energy, as shown in Fig. 5.8. Also, the FSF is totally excluding the AC coefficients from the calculations as shown in Eqs. (5.1 and 5.2).

Using the standard video benchmark frames in Appendix [A], our experimental results indicate that the AC coefficients have negligible effect on the direction decision computation, when using the non-homogeneity factor for such propose. Indeed, excluding the AC coefficients from our calculations dramatically reduces the Intra prediction over all computational complexity. On the other hand, the second proposed enhancement is in computing the non-homogeneity factor for the plane (P) direction as shown in Eq. (5.6). The following set of Eqs. (5.3–5.5) illustrates the non-homogeneity factors' calculations for each of the 16×16 macro-block's Intra reconstruction direction candidates:

$$H_{dc} = S \div 2 \tag{5.3}$$

$$H_v = |DC_0 - DC_2| + |DC_1 - DC_3| \tag{5.4}$$

$$H_h = |DC_0 - DC_1| + |DC_2 - DC_3| \tag{5.5}$$

$$H_p = |DC_0 - DC_3| + |DC_1 - DC_2| \tag{5.6}$$

Accordingly, after calculating the non-homogeneity factors for each of the 16×16 macro-block reconstruction direction candidates, the direction decision would be the direction that achieves the minimum non-homogeneity factor.

In this stage, The FSF algorithm nominates only one reconstruction direction for the Intra 16×16 mode without the need to execute any Rate Distortion Operations (RDO). In contrast, the H.264/AVC standard Intra direction prediction for the Luma

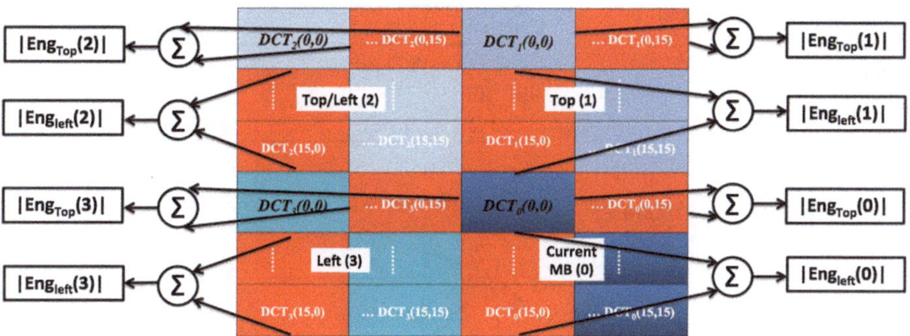

Fig. 5.9 The DCT coefficients used for (θ) calculations

16×16 mode executes one RDO for each of the four direction prediction candidates. Accordingly, the FSF eliminates all the required RDO computations form the Intra Luma 16×16 mode, which effectively reduces the computational complexity of the over all Intra prediction process.

On the other hand, in case of the Intra Luma mode decision is decided to be the 4×4 mode, the direction prediction process in the FSF algorithm computes the edge orientation factor as in Eq. (5.7). In fact, we have adopted the edge orientation technique introduced in [9]. However, we have significantly optimized the process of calculating the edge orientation as discussed in [15].

First, the angle (θ) is estimated from the DCT coefficients of the current 16×16 macro-block's samples and the neighboring 16×16 macro-block's samples: left, top and top/left, as shown in Fig. 5.9. The process of calculating the angle (θ) is modelled in Eq. (5.7). Also, the energy of the top and left of each of the four macro-blocks is calculated as in Eqs. (5.8 and 5.9).

$$\theta = arctan\left\langle \left\{ \sum_{m=0}^{3} Eng_{top}(m)/Eng_{left}(m) \right\} /4 \right\rangle; \tag{5.7}$$

where

$$Eng_{top}(m) = \sum_{i=0,j=1}^{j=15,j=j+1} |DCT_m(i,j)| \tag{5.8}$$

$$Eng_{left}(m) = \sum_{i=1,j=0}^{i=15,i=i+1} |DCT_m(i,j)| \tag{5.9}$$

Based on this angle (θ), the Luma 4×4 macro-block direction candidates are chosen from Table 5.2. Obviously, the direction candidate that achieves the minimum RD value is chosen as the reconstruction direction for the current 4×4 macro-block.

Table 5.2 The FSF direction candidates for the Intra 4 × 4 mode

Candidate group	Intra 4 × 4 candidates
60:90 OR −90:−60	0, 2, 5, 7
−30:30	1, 2, 6, 8
30:60 OR −60:−30	0, 1, 2, 3

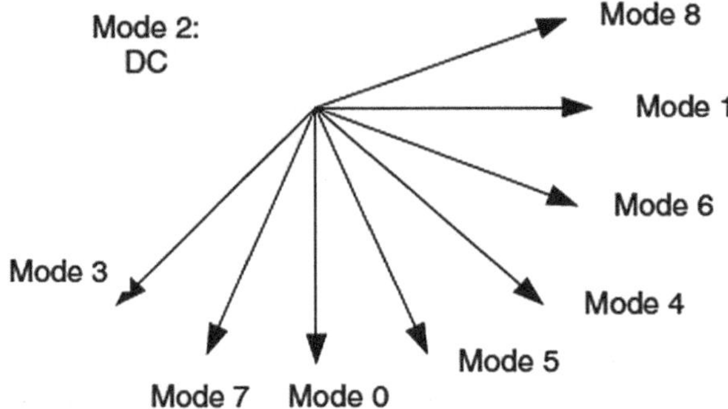

Fig. 5.10 The standard direction candidates for the Intra 4 × 4 mode

Clearly, the FSF algorithm nominates only four direction candidates for each of the 16 4 × 4 sub macro-block. Each of the 4 × 4 sub macro-block will need only four RDO computations to decide the reconstruction direction. In contrast, the standard H.264/AVC Intra 4 × 4 Luma mode calculates the RDO for each of the nine direction prediction candidates, which are shown in Fig. 5.10.

In brief, the FSF Intra prediction algorithm cuts down the required number of RDO from nine times in the standard Intra Luma 4 × 4 mode to only four RDO, which significantly reduces the computational complexity of the over all Intra prediction process. By deciding the 4 × 4 direction prediction, the FSF Intra prediction algorithm has efficiently concluded the Intra mode selection and reconstruction direction prediction for Luma component of the currently decoded macro-block.

On the other hand, the Chroma Intra prediction has only one mode in the H.264/AVC baseline profile. Moreover, the proposed FSF algorithm predicts the reconstruction direction for both Chroma components of the current 16 × 16 macro-block without performing any additional computations. The proposed algorithm only checks if the current Luma 16 × 16 direction prediction is the DC. And, in case if the current Luma 16 × 16 reconstruction direction is DC, then the Chroma reconstruction direction will be the DC direction. Otherwise, the Chroma reconstruction direction candidates will be the DC direction and the reconstruction direction of the current Luma 16 × 16 macro-block. In the worst case, only two RD operations will be needed to decide the reconstruction direction for both of the current macro-block Chroma components. That is saving 75 % of the required RDOs by the H.264/AVC standard Intra prediction.

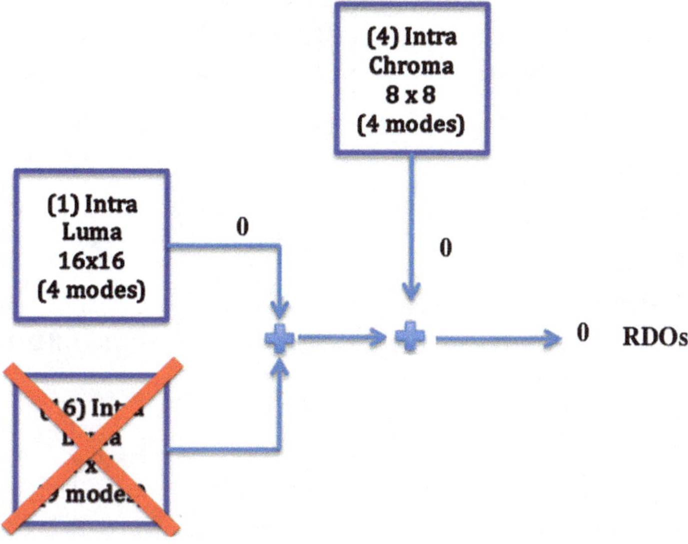

Fig. 5.11 The FSF required RDOs in best case

5.3 The FSF Intra Prediction Complexity Analysis

In the Intra prediction algorithm currently employed by the standard H.264/AVC, each macro-block's Luma component and its both Chroma components use the Rate Distortion (RD) to decide the current Intra macro-block mode and its reconstruction direction. Accordingly, the computational complexity of the Intra prediction algorithm can be estimated from the total number of Rate Distortion Operations (RDO) that the Intra prediction algorithm needs to execute in order to decide the Intra macro-block mode and the reconstruction direction for both the Luma and the Chrom components. In brief, the total number of RDO operations can be modelled as in Eq. (5.10).

$$RDO_{No} = M8 * (16 * M4 + M16) \tag{5.10}$$

where M8, M4 and M16 are the number of direction prediction candidates for the 8×8 Chroma mode, 4×4 Luma mode and 16×16 Luma mode, respectively. In fact, as shown in Fig. 2.4, for each 16×16 macro-block, the standard Intra prediction process has to perform a total of:

$4 * (16 * 9 + 4) = 592$ RDOs [19].

On the other hand, the FSF Intra prediction algorithm handles the Chroma and Luma components independently. Additionally, it succeeds to eliminate or reduce the required number of RDOs in each mode. As shown in Fig. 5.11, the total number of RDOs will be reduced to:

$0 + (16 * 0 + 0) = 0$ RDO in the best case.

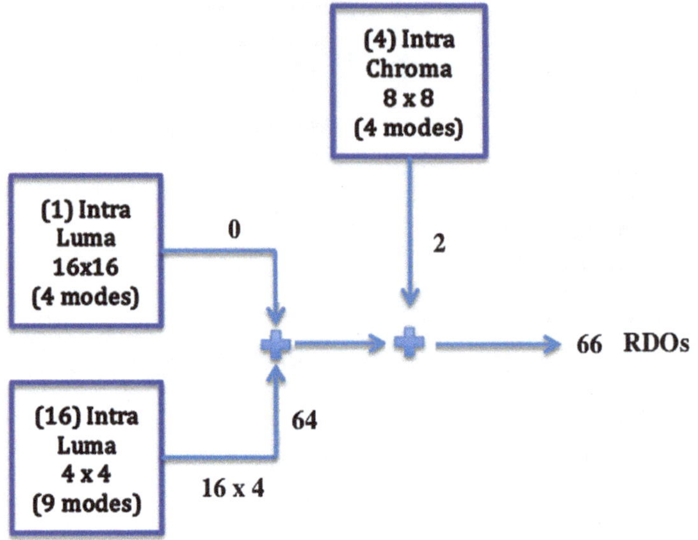

Fig. 5.12 The FSF required RDOs in worst case

Table 5.3 The required RDOs by the FSF versus the H.264/AVC standard

Macro-block mode	Standard H.264/AVC	FSF best case	FSF worst case
Luma 4×4	9	0	4
Luma 16×16	4	0	0
Chroma 8×8	4	0	2
Total	592	0	66

or,
$2 + (16 * 4 + 0) = 66$ RDOs in the worst case, as shown in Fig. 5.12.

Also, Table 5.3 compares the required number of RDO for the H.264/AVC standard and the FSF at the macro-block level.

In conclusion, the proposed FSF Intra prediction algorithm needs, in the worst case, only 11 % of the total number of RDO that the H.264/AVC standard algorithm computes to decide the Intra mode and reconstruction direction for each macro-block (Fig. 5.13).

5.4 The FSF Intra Prediction Software Implementation

For evaluation and comparison purposes, the proposed FSF Intra prediction algorithm and both Kim's algorithm in [8] and Yoo's algorithm in [9] have been implemented using C++ and integrated in the H.264/AVC reference software (JM 18.2).

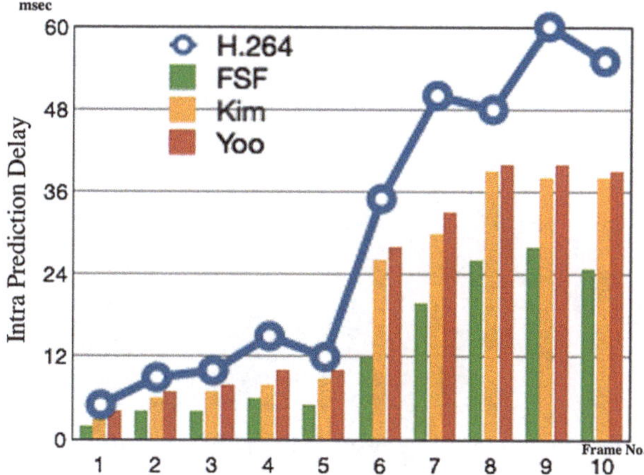

Fig. 5.13 The Intra prediction run-times for different benchmark frames

Table 5.4 The specifications of the system used for the experiments

Item	Tech specs
Processor	Intel Core 2 Due 2.4 GHz CPU
Memory	4 GB 1067 MHz DDR3 RAM
OS	Mac OS X Lion 10.7.2
Programming language	C++
Reference software	JM 18.2

Table 5.4 summarizes the technical specification of the system that has been used for implementing and testing all the previously mentioned algorithms.

Furthermore, the comparisons have been done in respect to the Intra prediction run-time, bit-rate and PSNR. The Intra prediction run-time reflects the computational complexity of the tested algorithm. While the bit-rate is a measure to the compression efficiency of the applied algorithm. Finally, the PSNR is used for subjective evaluation of the resulting video quality from the proposed FSF Intra prediction algorithm, the Intra prediction technique currently used in the standard H.264/AVC JM 18.2 and both of the enhanced Intra prediction techniques that have been introduced by Kim's algorithm [8] and Yoo's algorithm [9], respectively.

Successively, the empirical results show that our proposed FSF Intra prediction algorithm achieved a reduction in Intra prediction run-time of 56 % less than the Intra prediction run-time that is taken by the standard Intra prediction algorithm in H.264/AVC JM 18.2. On the other hand, the FSF Intra prediction is offering almost the same bit-rate as the standard H.264/AVC Intra prediction. Yet, the cost was a little drop in the PSNR by a factor of 1.8 % (or 0.72 dB in average) less than the PSNR that is achieved by the standard H.264/AVC Intra prediction algorithm.

Furthermore, both Kim's and Yoo's algorithms introduce some enhancements in the Intra prediction run-time with proportionally larger cost in the PSNR. Also, the FSF Intra prediction algorithm achieves in average 35.29 and 39.73 % reduction in

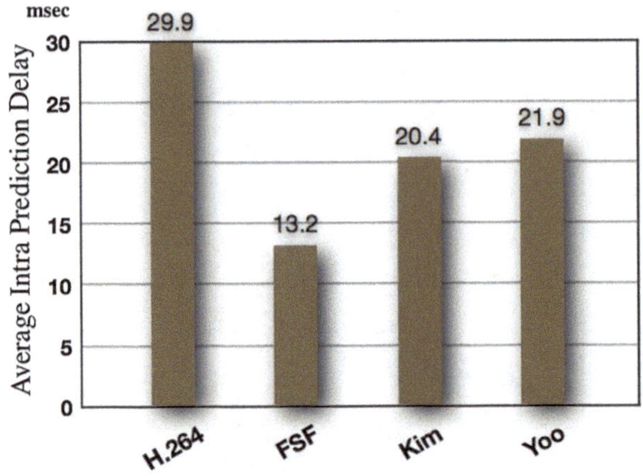

Fig. 5.14 The average achieved Intra prediction run-time

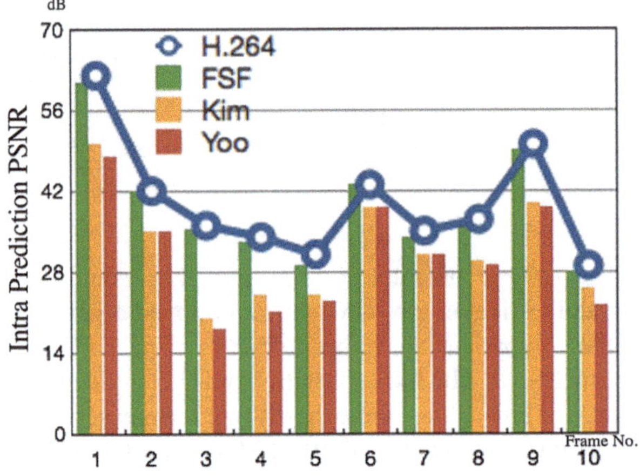

Fig. 5.15 The PSNR for different benchmark frames

the Intra prediction run-time when it is compared to the consumed run-time by each of Kim's and Yoo's Intra prediction techniques, respectively. In addition, the FSF Intra prediction algorithm shows an average improvement in the PSNR by 18.84 and 22.10 % more than the PSNR introduced by Kim's and Yoo's algorithms, respectively. Figures 5.16 and 5.14 illustrate all the previously mentioned results and comparisons.

On the other hand, Figs. 5.15 and 5.13 show the PSNR cost and the Intra prediction run-time that have been achieved by testing all the implemented algorithms on the used standard benchmark, which is composed of ten standard image and video processing frames of different sizes and textures. Table 5.5 maps the frame names to their numbers in Figs. 5.15 and 5.13. Also, a thumbnail figure of each of the tested

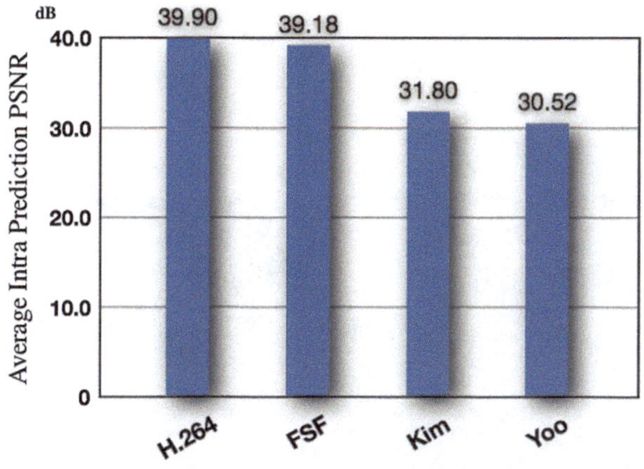

Fig. 5.16 The average achieved PSNR

Table 5.5 Benchmark frame name to number directory

Frame no	Frame name	Appendix
1	Cameraman	A.1
2	Foreman	A.2
3	House	A.3
4	Jet plane	A.4
5	Lake	A.5
6	Lena	A.6
7	Mandrill	A.7
8	Peppers	A.8
9	Pirate	A.9
10	Bridge	A.10

benchmark frames are presented in the Appendix [A] of this book. Finally, Fig. 5.17 is an example of the objective quality achieved by the proposed FSF Intra prediction, which is comparable to the objective quality of the standard H.264/AVC.

5.5 Efficient FSF Intra Prediction Hardware Architecture

The currently employed Intra prediction technique in the H.264/AVC standard decoder introduces a high degree of computational complexity that causes an unacceptable decoding delay especially for the real-time video applications. In contrast, the proposed FSF Intra prediction solution was able to reduce the Intra prediction run-time and qualify the H.264/AVC software decoder for the real-time video applications. However, the software solutions are by nature very dependent on the underlying

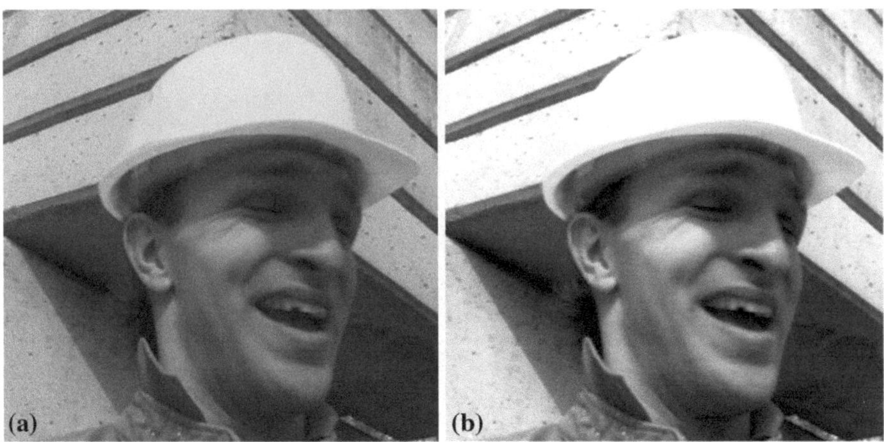

Fig. 5.17 *Foremen* using **a** standard and **b** FSF Intra prediction

hosting hardware. Even with powerful high-speed CPUs, a backlog still occurs. Most often, the information throughput capability of the software solution can not keep up with the minimum throughput requirement of the real-time video applications.

Here, the real benefit of the application specific hardware designs shine. Specially, when split-second situational awareness and decision making require the processing of huge volumes of information in real-time. Furthermore, the hardware-based solution introduces more reliability than the software solutions. Indeed, the need for speed necessitates a hardware solution. For that reason, we introduce in this section an efficient hardware architecture design for the FSF Intra prediction algorithm that we have discussed in details in Sect. 5.2 of this chapter.

All things considered, the design of the proposed FSF Intra prediction architecture is divided into five functionally integrated modules. In the first place, the current macro-block's DCT coefficients are applied to the first module of the proposed Intra FSF architecture directly from the standard H.264/AVC inverse quantization block (Q^{-1}). Thus, all the data is processed in the DCT domain. Hence, our early Intra prediction determining technique does not wait for the inverse DCT (T^{-1}) in the H.264/AVC decoder as shown in Fig. 5.18.

Specifically, only the DCT's DC coefficient of the current macro-block along with the DCT's DC coefficients of the neighboring three macro-blocks: the top, the top/left and the left are applied to the first FSF module, which is the Smoothness Factor (SF) module to estimate the value of the current macro-block smoothness. As shown in Fig. 5.19, the SF module simply consists of simple adders and shift units only.

Subsequently, as shown in Fig. 5.20, the output of the SF module is applied to the Mode Decision (MD) module. The MD module is composed of a simple comparator unit that compares the current output value of the SF module against a predetermined quantization based threshold (Th). Although the MD module architecture is very

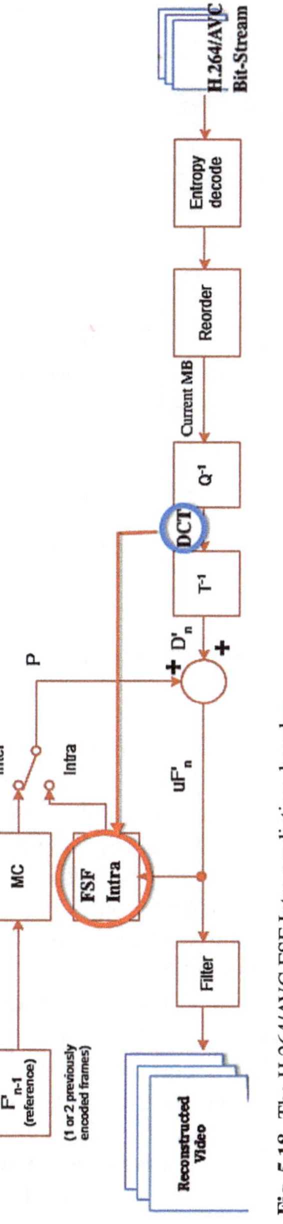

Fig. 5.18 The H.264/AVC FSF Intra prediction decoder

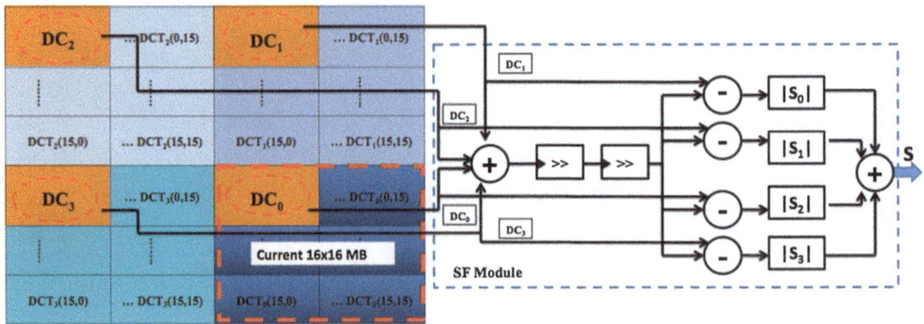

Fig. 5.19 The FSF smoothness factor (SF) module

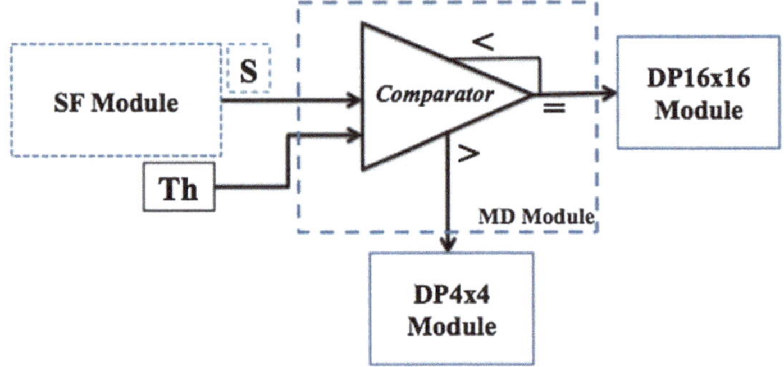

Fig. 5.20 The FSF mode decision (MD) module

simple, it is a quite critical stage that controls the Intra prediction mode selection for the current macro-block. Also, the threshold (Th) value is set ahead of time in order to fit the application requirements. In other words, some applications require a high video quality and can tolerate the resulting higher encoding delay or lower compression rate. In contrast, other applications can tolerate video quality in order to achieve a higher compression or better encoding speed. Generally, selecting the threshold value is an application dependent process [20].

Based on the output value of the MD module, either the 16×16 Direction Prediction (DP16 × 16) module or the 4×4 Direction Prediction (DP4 × 4) module will be activated. Such split architecture is introducing a significant reduction to the overall power consumption of the FSF Intra prediction hardware design. The shortest path in such split architecture is started by the DP16 × 16 module, which is activated only if the Intra Luma 16×16 mode has been selected by the MD module. The DP16 × 16 module reuses the smoothness factor (S), which has been previously calculated by the SF module, and the DC coefficients of the current macro-block and its neighbouring macro-blocks, as shown in Fig. 5.8. All these parameters are applied as inputs for Non-Homogeneity (NH) module to calculate the non-homogeneity factor. Figure 5.21 shows the NH module as a sub-module in the DP16 × 16 Module.

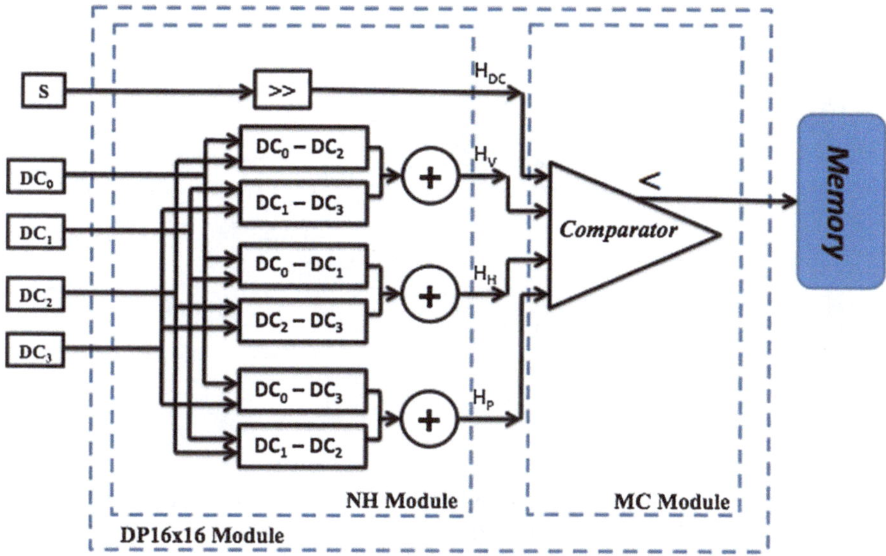

Fig. 5.21 The FSF Luma 16 × 16 direction prediction (DP16 × 16) module

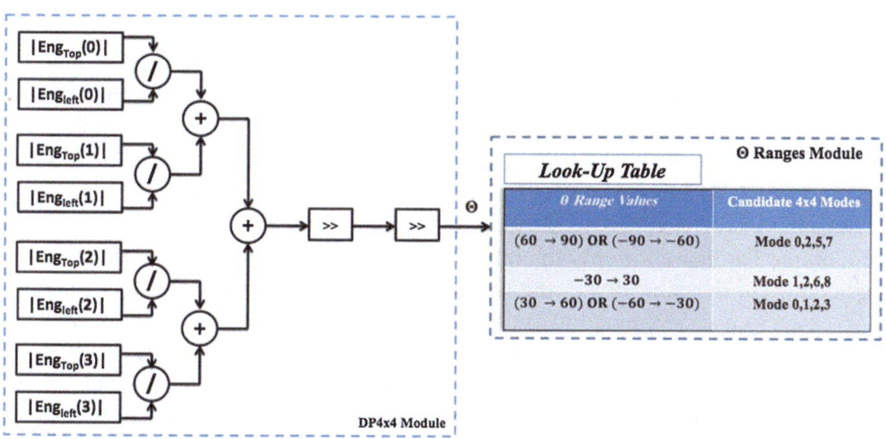

θ Range Values	Candidate 4x4 Modes
(60 → 90) OR (−90 → −60)	Mode 0,2,5,7
−30 → 30	Mode 1,2,6,8
(30 → 60) OR (−60 → −30)	Mode 0,1,2,3

Fig. 5.22 The FSF Luma 4 × 4 direction prediction (DP4 × 4) module

Furthermore, the non-homogeneity factors, which are the output value of the NH module, are applied to the Minimum Cost (MC) module. Then, the MC module, which is composed of a simple comparator unit, nominates the candidate direction for the whole 16 × 16 macro-block as one unit and saves it into the designated memory unit. All in all, the DP16 × 16 module represents the shortest path for the FSF Intra prediction split architecture and the operation is done in half the period.

Alternatively, the other possible path is through the DP4 × 4 module, which only gets activated if the Intra Luma 4 × 4 mode has been selected by the MD module.

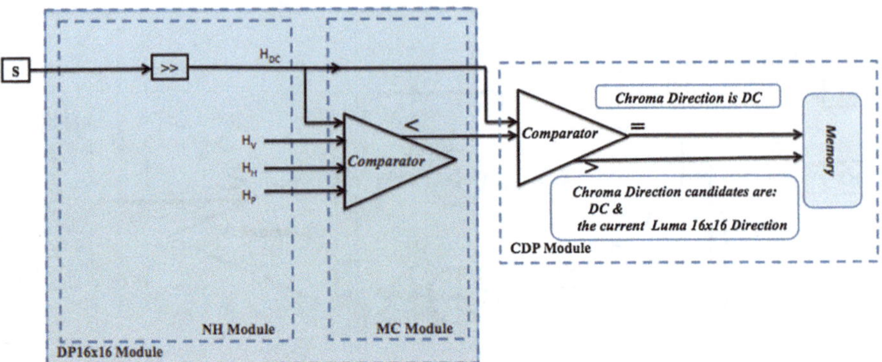

Fig. 5.23 The FSF Chroma direction prediction (CDP) module

In other words, the DP4 × 4 module will be awaken from the sleep state into the functional state by the output of the MD module. At the same time, the DP16 × 16 module will be moved to the sleep state. Figure 5.22 shows the architecture design of the proposed DP4 × 4 Module, which is used to compute the edge orientation for the FSF Intra prediction process at macro-block level.

At first, the top samples energy and the left samples energy are calculated for the current Luma 16 × 16 macro-block as shown in Fig. 5.9. Then, the accumulated energy values are applied as an input to the DP4 × 4 module. In essence, the DP4 × 4 Module consists of a parallel-array divider, which is designed by using simple small delay logic gates only. Furthermore, the computed angle (θ) is sent to the (θ) Ranges Module, which is composed of a simple look-up table as shown in Fig. 5.22.

In fact, the proposed FSF Intra prediction architecture until this point is describing the required hardware for only the Intra Luma mode and its corresponding direction prediction. On the other hand, the Chroma Direction Prediction (CDP) module is all we used to decide the reconstruction direction for both components of the 8 × 8 Chroma macro-blocks at the H.264/AVC baseline profile as illustrated in Fig. 5.23. Also, the CDP module is only composed of a simple comparator unit, which is used to decide the reconstruction direction candidates based on the currently selected direction for the corresponding Luma 16 × 16 component.

5.6 The FSF Intra Prediction Hardware Implementation

The H.264/AVC codec requires a significantly high computing power and specific types of arithmetic operations. However, the latest quad-core general-purpose x86 Central Processing Unit (CPU) has sufficient computation power to perform real-time Standard-Definition (SD) and High-Definition (HD) decoding. In addition, the CPU based solution offers much more flexibility, especially when the decoding must be done concurrently in different bit-rates or different resolutions on various display devices. In fact, the second generation Intel Core "i" processors, i.e. i3, i5 and i7, which have been introduced at the January 2011 Consumer Electronics Show (CES), offer an on-chip hardware full HD H.264/AVC codec [21]. The Intel marketing name for the on-chip H.264/AVC encoder feature is **"Intel Quick Sync Video"** [22].

Indeed, the compression efficiency is video algorithmic dependent. It is not affected by whether the hardware or the software is used for the video algorithm implementation. However, the software implementation that runs on general-purpose CPUs is typically less power efficient. Therefore, the main difference between hardware and software implementation is focusing more on power-efficiency, reliability and integration cost.

Accordingly, special-purpose hardware implementation, e.g. ASIC or FPGA, is strongly recommended to optimize the codec power efficiency. Either partial hardware acceleration or complete encoding or decoding modules may be employed to reduce the overall encoder or decoder power consumption, respectively. The H.264/AVC hardware codec can be implemented using either the Application Specific Integration Circuit (ASIC) or the Field-Programmable Gate Array (FPGA), which is simply a general programmable chip. Also, a partial or even a full H.264 encoder could run on a single low cost FPGA chip. However, H.264/AVC full codec design is required in order to customize the FPGA chip as a hardware codec for a specific video application or device.

In fact, both ASIC and FPGA H.264/AVC codecs are available from several semiconductor companies. However, the core design used in the ASIC is licensed by few companies, e.g. On2, which is managed by Google. Also, Texas Instruments (TI) manufactures a line of ARM + DSP cores that perform DSP H264/AVC BP encoding 1080p at 30 fps [23]. Such codecs are implemented in sophistically optimized DSP code to consume less power than the software codec running on generic CPU.

Accordingly, in the scope of this book, the FSF Intra prediction architecture has been prototyped on "Genesys Virtex-5 LTX FPGA". The FPGA implementation results are discussed in the following section. Even more, the encouraging results of the FSF's FPGA prototype motivate us to move the FSF Intra prediction algorithm into the ASIC implementation. In the final analysis, Sect. 5.6.2 discusses the results that we have obtained from the ASIC implementation of the proposed FSF Intra prediction architecture using the 45 nm technology.

5.6.1 The Virtex-5 FPGA Prototyping for FSF Intra Prediction

Generally speaking, the FPGAs are field programmable devices, which can be programmed and reprogrammed as required by the developer. Successively, as the developer programs his architecture design into the FPGA, the design boolean is fused within the FPGA logic gates. Although the cost per FPGA is expensive, a patent has to be given to the FPGA chip manufacturer. Also, the user has to choose a particular FPGA chip to match with the complexity of his design. For testing and evaluation

Fig. 5.24 The Genesys Virtex-5 LTX FPGA

purpose, the proposed FSF Intra prediction architecture has been implemented and tested on "Genesys Virtex-5 LTX FPGA" Development Board.

Figure 5.24 shows the board, which has been chosen for prototyping the proposed FSF Intra prediction architecture. The "Genesys" FPGA is equipped with "Virtex-5 LXT", which is integrated in 65n technology and optimized for high performance logic with advanced serial connectivity. In details, the Virtex-5 LXT includes 7,200 slices. Each slice is containing four 6-input LUTs, eight flip-flops with 1.7 M bits of fast block RAM and twelve digital clock managers, which nominate Genesys Virtex-5 board to be a perfect match for implementing the FSF Intra prediction architecture. In addition, the board includes HDMI video port with resolution up to 1600×1200 and 24-bit color. Besides the VHDC connectors, an additional 4 HDMI terminal-D connectors were added with Vmod MIB-VHDC Module Interface Board. Also, the Virtex-5 board includes 256 Mbyte DDR2 SODIMM with 64-bit wide data which makes it completely suitable for our experiment. However, the main reason behind choosing the Virtex-5 FPGA for the FSF Intra prediction prototyping was mainly its previously mentioned built-in features, i.e. high speed I/O, IP-core generator, automatic optimization routing for better performance, automatic clock trees and embedded CPU.

First of all, the FSF architecture design has been written in VHDL (Very high speed integrated circuits Hardware Description Language). Then, we compiled and simulated our design on Xilinx ISE Design Suite version 13.3 before we downloaded our proposed design on the Virtex-5 FPGA. The prototype hardware implementation for the proposed FSF Intra prediction architecture has been able to achieve an operating speed of 250 MHz. However, the average net power consumption is 112 mW, which is a Total Power of 692 mW with 567 mw of Leakage Power. Clearly, most of the power is consumed due to the internal leakage power of the FPGA. Table 5.6 summarizes the results that have been obtained from the FPGA implementation.

Table 5.6 The Genesys Virtex-5 FPGA implementation results

Parameter	Value
Board	Genesys Virtex-5 LTX
Design suite	Xilinx ISE version 13.3
Technology	65 nm
Power consumption	112 mW
Leakage power	567 mW
Total power	692 mW
Clock frequency	250 MHz

All in all, the achieved high operating speed nominates the proposed FSF algorithm and its hardware design for interactive real-time applications and devices. In fact, the reprogrammable board provides post simulation accurate results until the design is ready for the ASIC implementation. Our FPGA prototype testing results encourage me to extend our experimental implementation of the FSF Intra Prediction to the ASIC design. Also, the high power consumption introduced by the "Vertex-5" board, which is mostly due to the internal leakage power of the FPGA, was another motivation to extend the proposed hardware design to the ASIC implementation in order to gain more control over the chip's power consumption.

5.6.2 The 45 nm ASIC Implemantation for FSF Intra Prediction

Although the cost of the Application Specific Integration Circuit (ASIC) chip is highly expensive compared to the FPGA board, the cost per unit decreases in the case of mass production. In fact, the ASIC implementation usually has an operating speed proportional to the used technology. However, the ASIC main contribution is in the significant power consumption saving.

In contrast, the FPGA boards suffer a higher power consumption, which is usually due to the internal leakage power. For that reason, and to prove the functionality and the efficiency of the proposed FSF Intra prediction technique, we extended our work by introducing the ASIC design and implementation for the proposed FSF hardware architecture, which we have fully described in Sect. 5.5. Generally speaking, to implement a hardware architecture design on ASIC, the programmer first writes hardware description by Hardware Description Language (HDL). Then, he simulates the HDL to get the net list. Finally, he synthesizes the design net list to obtain the end product layout.

The proposed FSF Intra prediction architecture was first implemented using functional VHDL. Then, several simulations have been done to verify the behavior of our designs using the advanced simulation tool "ModelSim". Also, for testing purposes, we employed the testing benchmark in Table 5.5, which represents video frames with different sizes and textures. And, in order to process the frames' macro-block sequences, the Intra frames are divided into non-overlapped blocks of size 16×16

Table 5.7 The ASIC area of power net distribution synthesizing report

Layer name	Area of power net	Routable area	Percentage
Metal 1	417.6458	11785.0616	3.5439
Metal 2	0.0000	11785.0616	0.0000
Metal 3	0.0000	11785.0616	0.0000
Metal 4	0.0000	11785.0616	0.0000
Metal 5	1460.3000	11785.0616	12.3911
Metal 6	1412.8000	11785.0616	11.9881
Metal 7	0.0000	11785.0616	0.0000

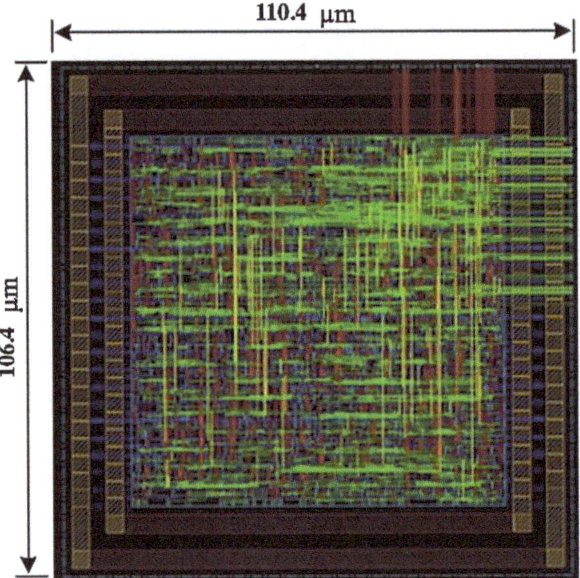

Fig. 5.25 The FSF Intra prediction 45 nm ASIC end product layout

using the "MATLAB Image Acquisition Toolbox". Last but not least, the proposed architecture has been implemented using TSMC 45 nm CMOS technology with seven layers of metal, but only three layers have dominant effect on the power consumption as shown in Table 5.7.

Furthermore, using "Cadence Ambit" synthesizer and "BGX-shell", the FSF architecture is synthesized, simulated, and analyzed. Moreover, the layout was done using "Cadence SOC Encounter" tool. The technology and the symbol libraries for synthesis and simulation in "Cadence" were generated based on Oklahoma State University (OSU) standard-cell library. Hence, Fig. 5.25 shows the obtained chip Layout of the FSF Intra prediction module using 45 nm technology.

Also, in order to demonstrate the size of the FSF Chip, Fig. 5.26 compares the dimension of the FSF Intra prediction chip to the dimension of the ASIC implementation of a complete H.264/AVC decoder in [24] which employs the standard Intra prediction. Clearly, integrating the proposed FSF Intra prediction chip into the stan-

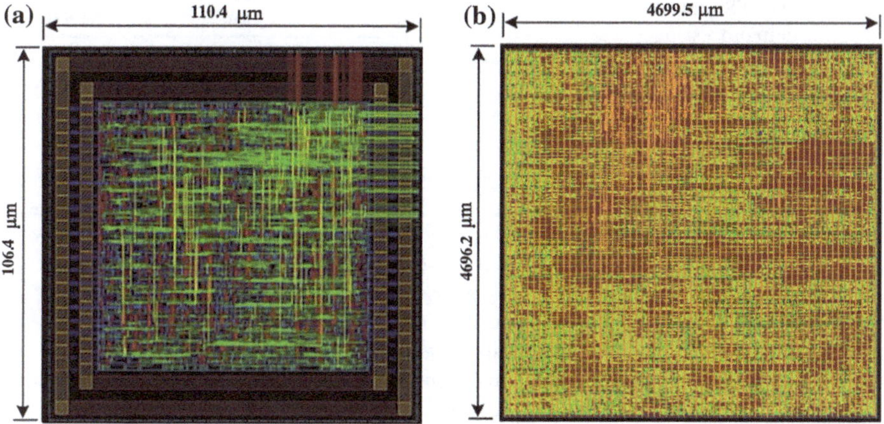

Fig. 5.26 ASIC layout of **a** FSF and **b** full standard H.264/AVC chips

dard H.264/AVC decoder adds a negligible increase (0.053 %) in the overall decoder chip size.

Also, the gate count, which can be defined as the number of transistor switches or gates needed to implement a design, is one of the most important factors that directly affects the end cost of the ASIC chip production. In other words, the less the number of gates required for a design, the cheaper the chip cost of that design will be. In particular, the gate count for the FSF Intra prediction ASIC design is 2,100 gates, which all are encapsulated in the $0.22\,\mathrm{mm}^2$ chip area. Moreover, the chip testing results show a stable operating frequency of 140 MHz. Accordingly, the produced FSF Intra prediction chip is recommended for all multimedia devices that favour the high throughput decoding at low-power consumption as a trade off to an negligible increase of 0.053 % in the overall chip area and preserving the same compression efficiency of the standard H.264/AVC decoder.

Furthermore, the hardware simulation ensured that our enhanced H.264/AVC Intra prediction decoder can easily perform real-time video processing. Specifically, when integrated in a whole H.264/AVC system, the FSF Intra prediction ASIC chip can perfectly perform the Intra prediction function for the H.264/AVC decoder at a frame rate of over 30 fps of 4CIF (704 × 576). Table 5.8 lists all the video sequences that have been tested on the FSF Intra prediction modified H.264/AVC decoder. Also, the output frames from our H.264/AVC system have the same visual quality of the original H.264/AVC system.

On the other hand, the power consumption sources in the proposed architecture can be classified into three main sources. First, the dynamic power also is well known as the switching power consumption. It can be defined as the power consumed when the transistors are switching during its actual logical functions. And, because of our split architecture design, only one half of the manipulation units will be consuming dynamic power at a time. Clearly, the switching power is not consumed by the devices that are in the sleeping state.

Table 5.8 The FSF Intra prediction chip processing capabilities

Video sequence	Resolution	Speed (fps)
Flower	QCIF 176 × 144	430
Football	CIF 352 × 288	110
Foreman	4CIF 704 × 576	30

Table 5.9 The FSF ASIC synthesizing summary results

Parameter	Value
Technology	TSMC 45 nm CMOS
Power consumption	9.01 mW
Leakage power	1.00 mW
Core chip area	0.118 mm^2
Total chip area	0.221 mm^2
Chip dimensions	110.4 × 106.4 μm
Gate count	2,100
On-chip memory	8k bit
Clock frequency	140 MHz

Table 5.10 The ASIC versus FPGA implementation results

Parameter	ASIC	FPGA
Technology	45 nm	65 nm
Power consumption	9.01 mW	692 mW
Leakage power	1.00 mW	567 mW
Clock frequency	140 MHz	250 MHz

The second source of power consumption is the leakage power consumption. It results from the current, which leaks through the transistors, even when the circuit is not being used. Lastly, the internal power consumption that includes the short circuit power.

Table 5.9 summarizes the results, which have been achieved from the proposed 45 nm ASIC implementation for the FSF Intra prediction architecture. In brief, the FSF Intra prediction ASIC design introduces a synthesized chip area of almost 0.22 mm^2. However, its overall power consumption is only 9.01 mW, which includes a leakage power of 1 mW. The chip low-power behaviour was a result to many optimization techniques that have been employed in our design, e.g. using only simple small delay logic gates and the split architecture technique, which makes only one half of FSF structure be used at a time as explained in Sect. 5.5. In addition, it enables the proposed FSF 45 nm ASIC chip design to operate stably at 140 MHz frequency. Indeed, such implementation results nominate the proposed FSF Intra prediction architecture for interactive real-time multimedia mobile devices.

In fact, comparing the FSF Intra prediction FPGA prototype to its ASCI implementation is not a fair comparison. As each of the two hardware implantations is using a different technology. While the FSF Intra prediction ASIC implementation is employing the 45 nm CMOS technology, the Virtex-5 FPGA Prototype is manufactured with 65 nm technology. However, Table 5.10 is introduced only to highlight the privileges that motivate us to upgrade the proposed FSF Intra prediction architecture from the FPGA prototype to ASIC chip design.

5.7 Conclusion

The Intra mode decision and direction prediction process dramatically increases the complexity of the H.264/AVC decoder. For that reason, in this chapter, the properties of the DCT coefficients are employed to significantly reduce the overall run-time of the H.264/AVC decoder. Specifically, a high throughput early Intra mode decision and direction prediction algorithm is presented, which utilizes the DCT coefficients from the output of the inverse quantization module for such purpose and before it gets transformed back to the pixel domain.

Also, the proposed FSF algorithm selects the macro-block Intra mode then predicts the reconstruction direction in a completely full-search free technique. The experimental results show that the FSF Intra prediction achieves over 56 % reduction in the Intra prediction run-time while preserving the same bit-rate as the standard H.264/AVC. However, the cost of such run-time reduction is a negligible degradation of less than 1.8 % in PSNR compared to the PSNR achieved by Intra prediction algorithm currently applied in the H.264/AVC JM 18.2 reference software.

In addition, the work has been extended by introducing the low-power hardware architecture design for the proposed high throughput FSF Intra prediction algorithm. Furthermore, both the FPGA prototype and the 45 nm ASIC chip design implementation are developed for the proposed FSF Intra prediction hardware architecture. In closing, the proposed 45 nm ASIC design achieves a stable operating frequency of 140 MHz, which strongly nominates the FSF Intra prediction hardware design for real-time H.264/AVC decoder devices. Furthermore, the total power consumption of the ASCI design was only 9.1 mW, which qualifies the proposed chip design for low power mobile video applications.

Appendix A
Benchmarks

See Figs. A.1, A.2, A.3, A.4, A.5, A.6, A.7, A.8, A.9, A.10 and A.11.

Fig. A.1 Cameraman

T. Elarabi et al., *Real-Time Heterogeneous Video Transcoding for Low-Power Applications*, DOI: 10.1007/978-3-319-06071-2,

Fig. A.2 Foreman

Fig. A.3 House

Fig. A.4 Jetplane

Fig. A.5 Lake

Fig. A.6 Lena

Fig. A.7 Mandrill

Fig. A.8 Peppers

Fig. A.9 Pirate

Fig. A.10 Bridge

Fig. A.11 Blond

References

1. Pierre Larbier, in *AVC/H.264 encoding*, Broadcast Engineering. (2009), http://broadcastengineering.com/hdtv/avch-encoding/index3.html
2. Huang-Chih Kuo, Li-Cian Wu, Hao-Ting Huang, Sheng-Tsung Hsu, Youn-Long Lin. IEEE Transactions on Very Large Scale Integration (VLSI) Systems. *A Low-Power High-Performance H.264/AVC Intra-Frame Encoder for 1080p HD Video*, **19**,6, 925–938, doi:10.1109/TVLSI.2010.2045402. ISSN=1063-8210
3. Ostermann, J. and Bormans, J. and List, P. and Marpe, D. and Narroschke, M. and Pereira, F. and Stockhammer, T. and Wedi, T. (2004). IEEE Circuits and Systems Magazine. *Video coding with H.264/AVC: tools, performance, and complexity*, **4**,1, 7–28, doi:10.1109/MCAS.2004.1286980. ISSN=1531-636X
4. Al, A. and Rao, B.P. and Kudva, S.S. and Babu, S. and Sumam, D. and Rao, A.V. (2004). ICIP '04. 2004 International Conference on Image Processing. *Quality and complexity comparison of H.264 Intra mode with JPEG2000 and JPEG*, **1**, 525–528, doi:10.1109/ICIP.2004.1418806. ISSN=1522-4880
5. Shaaban, M. and Bayoumi, M. (2009). IEEE International Symposium on Multimedia, 2009. ISM '09. 11th *A Low Complexity Inter Mode Decision for MPEG-2 to H.264/AVC Video Transcoding in Mobile, Environments*, 385–391, doi:10.1109/ISM.2009.39
6. Yoonjeong Shin and Namrye Son and Nguyen Dinh Toan and Gueesang Lee (2010). International Conference on Information Science and Applications (ICISA). 11th *Low-Complexity Heterogeneous Video Transcoding by Motion Vector Clustering*, 1–6, doi:10.1109/ICISA.2010.5480281
7. Richardson, I.E. : The H.264 Advanced Video Compression Standard. John Wiley & Sons (2010)
8. Xingang Liu and Kook-Yeol Yoo and Sung Won Kim (2010). IEEE Transactions on Consumer Electronics. *Low complexity intra prediction algorithm for MPEG-2 to H.264/AVC transcoder*, **56**,2, 987–994, doi:10.1109/TCE.2010.5506030. ISSN=0098-3063
9. Liu, Xingang and Yoo, Kook-Yeol (2007). International Conference on Intelligent Pervasive Computing (IPC). A *Fast Intra MB Mode Decision Method for the MPEG-2 to H.264 Transcoder*, 19–22, doi:10.1109/IPC.2007.47
10. *H.264/MPEG-4 Part 10 White Paper*. (2003), http://vcodex.com
11. Wiegand, T. and Schwarz, H. and Joch, A. and Kossentini, F. and Sullivan, G.J. (2003). IEEE Transactions on Circuits and Systems for Video Technology. *Rate-constrained coder control and comparison of video coding standards*, **13**,7, 688–703, doi:10.1109/TCSVT.2003.815168. ISSN=1051-8215

T. Elarabi et al., *Real-Time Heterogeneous Video Transcoding for Low-Power Applications*, DOI: 10.1007/978-3-319-06071-2,

12. Yu Liu and Wei Lu and Li Wang and Kaihua Liu (2006). EEE Tenth International Symposium onConsumer Electronics, ISCE. *Design of MPEG-2 to H. 264/AVC Transcoder*, 1–3, doi:10.1109/ISCE.2006.1689449
13. Li Wang and Qi Wang and Yu Liu and Wei Lu (2006). IEEE Tenth International Symposium on Consumer Electronics, ISCE '06. *A fast Intra Mode Decision Algorithm for MPEG-2 to H.264 Video Transcoding*, 1–5, doi:10.1109/ISCE.2006.1689477
14. Kalva, H. and Petljanski, B. (2006). IEEE Transactions on Consumer Electronics. *Exploiting the directional features in MPEG-2 for H.264 intra transcoding*, **52**,2, 706–711, doi:10.1109/TCE.2006.1649701. ISSN=0098-3063
15. Elarabi, T.A. and Ragab, A.M. and Mahmoud, H. and Bayoumi, M. (2011). IEEE Workshop on Signal Processing Systems (SiPS). *High speed intra mode and direction prediction for MPEG-2 to H.264/AVC realtime transcoder*, 78–83, doi:10.1109/SiPS.2011.6088953. ISSN=2162-3562
16. Shaaban Mohsen and M. Bayoumi (2009). ProQuest Dissertations and Theses. *Low complexity paradigm for heterogeneous video transcoding in mobile environments*, **13**,7, 688–703. ISSN=9781109621211
17. Yu-Wen Huang and Bing-Yu Hsieh and Tung-Chien Chen and Liang-Gee Chen. (2005). IEEE Transactions on Circuits and Systems for Video Technology. *Analysis, fast algorithm, and VLSI architecture design for H.264/AVC intra frame coder*, **15**,3, 378–401, doi:10.1109/TCSVT.2004.842620. ISSN=1051-8215
18. Yu-Kun Lin and Chun-Wei Ku and De-Wei Li and Tian-Sheuan Chang. (2009). IEEE Transactions on Circuits and Systems for Video Technology. *A 140-MHz 94 K Gates HD1080p 30-Frames/s Intra-Only Profile H.264 Encoder*, **19**,3, 432–436, doi:10.1109/TCSVT.2009.2013511. ISSN=1051-8215
19. Zhenyu Wei and Hongliang Li and King Ngi Ngan. (2007). IEEE International Symposium on Circuits and Systems, ISCAS 2007. *An Efficient Intra Mode Selection Algorithm For H.264 Based On Fast Edge Classification*, 3630–3633, doi:10.1109/ISCAS.2007.378539
20. Chang, S.G. and Bin Yu and Vetterli, M. (2000). IEEE Transactions on Image Processing. *Adaptive wavelet thresholding for image denoising and compression*, **9**,9, 1532–1546, doi:10.1109/83.862633. ISSN=1057-7149
21. *Quick Reference Guide to generation Intel Core Processor Built-in Visuals-Intel Software Network.* (2011), http://software.intel.com/en-us/articles/quick-referenceguide-to-intel-integrated-graphics/
22. Pierre Larbier, in *Intel Quick Sync Video.* (2010), http://www.intel.com/content/www/us/en/architecture-and-technology/quick-sync-video/quick-sync-video-general.html
23. http://processors.wiki.ti.com/index.php/Category:DM6467. (2011), http://broadcastengineering.com/hdtv/avch-encoding/index3.html
24. Ismail, Y. and Bayoumi, M.A. (2010). IEEE Workshop on Signal Processing Systems (SIPS). *Efficient high speed lattice-CORDIC IFFT architecture for DMT transmitter*, 151–155, doi:10.1109/SIPS.2010.5624779. ISSN=1520-6130

Zeitfracht Medien GmbH
Ferdinand-Jühlke-Straße 7
99095 Erfurt, Deutschland
produktsicherheit@kolibri360.de